T0297353

Medical Innovation

Medical Innovation

Concept to Commercialization

Edited by

Kevin E. Behrns
Vice President Medical Affairs, Dean School of Medicine,
Saint Louis University, St. Louis, MO, United States

Bruce Gingles
Vice President, Global Technology Assessment and Healthcare Policy,
Cook Medical Holdings, LLC, Bloomington, IN, United States

Michael G. Sarr
Professor Emeritus, Mayo Clinic, Rochester, MN, United States

Academic Press is an imprint of Elsevier
125 London Wall, London EC2Y 5AS, United Kingdom
525 B Street, Suite 1800, San Diego, CA 92101-4495, United States
50 Hampshire Street, 5th Floor, Cambridge, MA 02139, United States
The Boulevard, Langford Lane, Kidlington, Oxford OX5 1GB, United Kingdom

Notices
Knowledge and best practice in this field are constantly changing. As new research and experience broaden our understanding, changes in research methods, professional practices, or medical treatment may become necessary.

Practitioners and researchers must always rely on their own experience and knowledge in evaluating and using any information, methods, compounds, or experiments described herein. In using such information or methods they should be mindful of their own safety and the safety of others, including parties for whom they have a professional responsibility.

To the fullest extent of the law, neither the Publisher nor the authors, contributors, or editors, assume any liability for any injury and/or damage to persons or property as a matter of products liability, negligence or otherwise, or from any use or operation of any methods, products, instructions, or ideas contained in the material herein.

Library of Congress Cataloging-in-Publication Data
A catalog record for this book is available from the Library of Congress

British Library Cataloguing-in-Publication Data
A catalogue record for this book is available from the British Library

ISBN: 978-0-12-814926-3

For information on all Academic Press publications visit our website at https://www.elsevier.com/books-and-journals

Publisher: John Fedor
Acquisition Editor: Mary Preap
Development Editor: Jennifer Horigan
Production Project Manager: Poulouse Joseph
Designer: Miles Hitchen

Typeset by SPi Global, India

Dedication

To Bill Cook, entrepreneur, industrialist, philanthropist, faithful friend to the medical profession, and tireless patient advocate. Bill made his engineering and manufacturing talents accessible to inventive physicians around the world, beginning with Charles Dotter, MD. Together, they ushered in what would later become known as the field of interventional radiology. Their partnership became a model for innovative medical approaches that would follow, including interventional cardiology and therapeutic gastrointestinal endoscopy. "Ready, fire, aim!"

Contents

Contributors

Numbers in parentheses indicate the pages on which the authors' contributions begin.

Dan E. Azagury (117), Stanford University, Stanford, CA, United States

Kevin E. Behrns (221), Saint Louis University School of Medicine, St. Louis, MO, United States

Katherine S. Blevins (117), Stanford University, Stanford, CA, United States

Mark Boden (51), Boston Scientific Corporation, Quincy, MA, United States

David E. Chadwick (19), Cook Medical, Bloomington, IN, United States

Venita Chandra (117), Stanford University, Stanford, CA, United States

Mark S. Cohen (209), University of Michigan, Ann Arbor, MI, United States

Elizabeth Cole (109), Tahoe Institute for Rural Health Research, Truckee, CA, United States

Frank S. David (167), Pharmagellan LLC, Milton, MA, United States

Lyn Denend (251), Stanford School of Medicine; Stanford Byers Center for Biodesign, Stanford, CA, United States

Ryan D. Egeland (167), Medtronic, Plymouth, MN, United States

James C. Eisenach (231), Foundation for Anesthesia Education and Research, Schaumburg, IL, United States

Thomas J. Filarski (43), Steptoe & Johnson LLP, Washington, DC, United States

C. Corbin Frye (179), General Surgery Resident, Washington University School of Medicine, St. Louis, MO, United States

Johnathan Goldstein (51), Boston Scientific Corporation, Quincy, MA, United States

Andy Hayden (221), Saint Louis University School of Medicine, St. Louis, MO, United States

Theodore W. Heise (101), Cook Incorporated, Bloomington, IN, United States

Thomas Hobday (109), Tahoe Institute for Rural Health Research, Truckee, CA, United States

James Hood (109), Tahoe Institute for Rural Health Research, Truckee, CA, United States

Seth Klapman (209), University of Michigan, Ann Arbor, MI, United States

Thomas M. Krummel (117), Stanford University, Stanford, CA, United States

Priya Kumthekar (201), Northwestern University Feinberg School of Medicine, Chicago, IL, United States

Eric Leuthardt (129), Washington University, St. Louis, MO, United States

Thomas R. Mackie (129), University of Wisconsin, Madison, WI, United States

Josh Makower (251), Stanford School of Medicine; Stanford Byers Center for Biodesign, Stanford, CA; ExploraMed Development LLC, Mountain View, CA; New Enterprise Associates, Chevy Chase, MD, United States

Dennis Matthews (109), Tahoe Institute for Rural Health Research, Truckee, CA, United States

Rudy A. Mazzocchi (77), ELENZA, Inc., Roanoke, VA/Zürich, Switzerland

Kyle Miller (191), Bold Diagnostics, LLC, Chicago, IL; Intuitive Surgical Inc., Sunnyvale, CA, United States

Barry S. Myers (65), Department of Biomedical Engineering, Director of Innovation, CTSI, Coulter Program Director

Deepa Narayanan (87), SBIR Development Center, National Cancer Institute, Rockville, MD, United States

Jenell Paul-Robinson (147), MedTech Innovator, Los Angeles, CA, United States

Zachary Rapp (167), Pharmagellan LLC, Milton, MA, United States

Charles Robert Hallford (65), New Ventures, Duke University, Office of Licensing and Ventures; Adjunct Associate Professor of Finance, Fuqua School of Business, Durham, NC, United States

Lee L. Sanström (235), Institute for Image Guided Therapies, University of Strasbourg, Strasbourg, France; The Oregon Clinic, Portland, OR, United States

Michael G. Sarr (243), Mayo Clinic, Rochester, MN, United States

Jay Schrankler (31), University of Minnesota, Minneapolis, MN, United States

Thomas P. Stossel (1, 139), BioAegis Therapeutics, Inc., Belmont; Options for Children in Zambia, Foxboro, MA, United States

Mamta Swaroop (201), Northwestern University Feinberg School of Medicine, Chicago, IL, United States

William D. Voorhees III (101), MED Institute Inc., West Lafayette, IN, United States

James K. Wall (117), Stanford University, Stanford, CA, United States

Tom Walsh (9), Partner, Ice Miller LLP, Indianapolis, IN, United States

Michael Weingarten (87), SBIR Development Center, National Cancer Institute, Rockville, MD, United States

Elisabeth K. Wynne (117), Washington University School of Medicine, St. Louis, MO, United States

Carla L. Zema (155), Vice President, Precision Health Economics, Los Angeles, CA, United States

Preface

This book is a primer for those in many disciplines interested in improving health care through medical technology innovation. The intended readership comprises aspiring entrepreneurs, especially physicians and medical students, as well as leaders in the fields of translational research, medical education, technology transfer, hospital and medical school administration, and health care policymaking. The book was born out of a themed innovation series in the journal SURGERY that presented 24 topical articles between November 2016 and November 2017. This book describes technical, legal, regulatory, business, education, economic, and policy predicates that enable technology innovation in medicine and health care broadly. The material is generally presented in chronological order from invention conception through the various essential gates that transform novel ideas into tangible and useful products in clinical settings. The chapters combine professional information along with horizon scanning and specific lessons. We have made a concerted effort to expand beyond purely technical knowledge and to present perspectives not normally associated with developing a medical innovation. Workforce development, federal and institutional policy, the medical school curriculum, and technology adoption are key among them. With the exception of two recent medical school and residency graduates, the authors are nationally recognized experts, selected for high career achievement and their practical rather than theoretical orientation. Each author has directly nourished innovation and for most, through dozens of cycles.

We hope this book accelerates the reader's path from a curious problem solver to a successful entrepreneur, with patients as the ultimate beneficiary.

Bruce Gingles
Kevin E. Behrns
Michael G. Sarr

Chapter 1

Introduction to Medical Innovation

Thomas P. Stossel
BioAegis Therapeutics, Inc., Belmont, MA, United States

The purpose of this book is to serve as a users' manual for aspiring physician-entrepreneurs and other professionals who want to exert their energy toward improving health care by advancing its technology. As the readers will discover, the health care professional schools in which they are enrolled or from which they have graduated hardly impart most of the know-how required to achieve such a goal. This book attempts to remedy that deficiency, and this introductory chapter provides some historic background concerning health care innovation (defined narrowly as discoveries, inventions, or devices that benefit patients) to explain why this book is necessary.

Throughout most of recorded history—excepting the construction of instruments for removing teeth, amputating limbs, trephining skulls, and performing rudimentary surgery—little innovation in health care took place. Indeed, the adage falsely ascribed to Hippocrates "Do no harm" notwithstanding, devices for bleeding and nostrums for purging patients reliably did more harm than good, an outcome masked by spontaneous recoveries and the desperation of the sick.

Discoveries much celebrated by medical historians concerning the circulation of blood, anatomic structure, observations using microscopes, or the invention of the stethoscope contributed to the diagnosis of disease but had minimal impact on clinical outcomes. One exception, smallpox vaccination, was controversial for a long time despite eventually eradicating the disease.

Even as innovation in health care germinated in the 19th century, the application of advances we consider truly important today, such as anesthesia, antisepsis, and the germ theory of disease that informed and amplified vaccination against infection, had to overcome resistance by the public as well as by the professionals involved in health care at the time. For example, dexterous surgeons able to complete procedures rapidly to minimize the agonizing pain suffered by their patients scoffed at anesthesia considering it a crutch for their less-skilled counterparts. Also, many religious scholars of the time believed

Medical Innovation. https://doi.org/10.1016/B978-0-12-814926-3.00001-2

that pain was God's punishment and that anesthesia should not be developed. Considering that their abysmal track record relegated health care professionals—such as they were—to the dregs of social status, the fact that health entrepreneurship centered mainly on quackery is hardly surprising [1].

These circumstances changed in the latter half of the 19th century, as discoveries of chemistry, microbiology, and physiology—predominantly occurring in Western Europe—transformed medical schools into bastions for health care research and accelerated the appearance of drugs and medical devices that began to supplant improved nutrition, sanitation, and housing as modalities for increasing longevity and improving the quality of life.

As similar institutions evolved in North America over the ensuing years, their innovators relied heavily on the European precedents. Many physicians and surgeons routinely obtained training in European medical centers, although they often amplified what they learned. When the Mayo brothers, founders of the famed clinic in Minnesota, learned a new surgical procedure in, say, Vienna, they would return the following year having performed many more of the operations at their American headquarters in the interval than had ever been done where the procedure originated. The Doctors Mayo and many other American surgeons invented surgical devices that bear their names to this day.

The 1919 Flexner Report served to legitimize health care and catalyzed the concentration of American health care education into universities, made science a requirement for legitimacy of health care training, and raised organized allopathic health care to a dominant position. Although only a minority of health care professionals actually engaged in research, this institutionalization of American health care and the establishment of professional societies facilitated innovation by elevating its social status and providing venues for communication about innovations and recognition of its important contributions [2].

During this era, researchers tended to focus on solving immediate problems in health care. Specific innovations that resulted from this effort have their unique and idiosyncratic histories, but one can generalize that most arose from astutely observing sick patients or experimental work in animals, leading to hypotheses on which to base clinical interventions. Another generalization is that innovations appear iteratively, because the solution to one problem informs approaches to solving others [3].

One example of the former generalization was the work of George Whipple at the University of Rochester who in the early decades of the 20th century addressed the clinical problem of red blood cell deficiency—anemia. He made dogs anemic by bleeding them and tested the effect of feeding them different diets on their recovery. When he found that raw liver was the most effective treatment, he had patients with the then prevalent and often fatal condition of pernicious anemia ingest large amounts of liver, and he indeed observed clinical improvement. George Minot and William Murphy working at Boston's Peter Bent Brigham Hospital subsequently documented that liver extracts were even more therapeutically effective.

Although Whipple, Minot, and Murphy received the 1934 Nobel Prize for their accomplishments, their story illustrates how the practical outcome—a treatment for a disease—does not necessarily result from what the researchers believe to be the underlying mechanism. Minot and Murphy credited iron for the therapeutic effect of liver, which is not true. The observed improvement in anemia actually resulted from high concentrations of folic acid in the liver, and subsequent research revealed that pernicious anemia results from a deficiency of vitamin B_{12}. Folic acid can partially overcome this deficit, because folic acid is a component of a biochemical pathway related to vitamin B_{12}, but interestingly and unfortunately, giving folic acid alone actually makes another manifestation of vitamin B_{12} deficiency—the neurologic complications—worse. The elucidation of vitamin B_{12} deficiency as the true basis of pernicious anemia was only achieved by the 1950s. Many innovations of this era, such as X-ray imaging or aspirin for fever, came on line with no idea as to how they worked, illustrating the discrepancy between effective treatment and mechanistic understanding.

An example of the other generalization—how therapies evolve from one problem-solving challenge to another—started with the invention in 1929 of a urinary catheter by the Boston surgeon Frederic Foley. The "Foley catheter" is a device consisting of a flexible, hollow tube with a very pliable tip that includes an external balloon incorporated into the wall of the tube that can be expanded by injection of air after insertion of the catheter into the bladder via the urethra. Initially devised to impart local pressure to stop bleeding after prostatic surgery, it came into wide use to relieve diverse causes of lower urinary tract obstruction by serving as an outlet for continuous bladder decompression. Building on that principle, Thomas Fogarty developed a similar catheter (the Fogarty catheter) in 1961 that could be inserted into a blood vessel (vein or artery) with a clot upstream or downstream and threaded past the blood clot, its external balloon at its tip then inflated, and then the catheter retracted to remove the clot. The next iteration of this principle was the application of a balloon catheter by Andreas Grüntzig in the 1970s to dilate critically narrowed coronary arteries to restore adequate blood flow to the myocardium supplied by that artery. The proof of principle that such arteries are amenable to safe instrumentation led to the development of intraluminal coronary stents and subsequently the drug-eluting varieties that minimize thrombosis within the stent.

During this era, basic science research enriched knowledge concerning biochemical pathways and the immune system involved in certain disease processes that ultimately bore on health. Facilitating such linkage was the study of patients with what became recognized as genetically based disorders—that is, "experiments of nature." In a few of these cases, it was possible to achieve therapeutic benefits. For example, the understanding that patients who inherited one of many dysfunctional metabolic pathways that converted chemicals in certain foods into toxins enabled dietary measures that avoided this consequence. Appreciation that other patients were unable to synthesize certain types of antibodies against infection led to the therapeutic infusion of antibody transfusions and improved infection control.

One of the few benefits of armed conflicts and other catastrophes is that the injuries inflicted by war uncover new problems that teach improvements in clinical management and introduce new coping technologies. World War II was no exception and by necessity, served to elicit practical approaches to pain control, imaginative ideas in the treatment of trauma, and both supportive care and new therapies for infections endemic in the war theater. Military physicians working in the field or in government laboratories spearheaded these accomplishments.

Also during this time, health care innovators were not considered "physician-entrepreneurs." In fact, the culture of the health care professional had long declared medical practice as a sacred calling as part of a—largely unsuccessful—effort to prop up its low status. This theologic imperative spilled over into health care research in universities and research institutes that frowned actively on researchers obtaining patents and profiting from their discoveries. But, as therapeutic opportunities accumulated and needed manufacturers to develop and improve innovators' responses to them, bringing drugs and medical devices to the market and to the patient, patents were required in order for development to proceed. Given that infectious disease was the leading challenge for health care in the first half of the 20th century, the introduction of antibiotic therapy not only placed the innovators of new, clinically effective antibiotics into partnerships with industry but also stimulated industry's interest and involvement in health care innovation. Prime examples included the collaboration of Paul Ehrlich with the I.G. Farben dyestuffs company leading to the discovery of Salvarsan, Kurt Domagk's association with Bayer to develop sulfonamides, Howard Flory and Ernst Chain's work with ICI and subsequent efforts at Pfizer that enabled mass production of penicillin, and Selman Waksman's connection with Merck that resulted in the availability of streptomycin.

The post-World War II period ushered in a huge uptick in federal funding for biomedical research. This largesse eventuated not only in the construction of the world's largest biomedical research facility, the NIH "intramural" facility in Bethesda, Maryland but also stimulated a marked increase in research grant awards to universities and independent research institutes that enabled these institutions to expand their research activities.

Without question, the seven decades that have transpired since the onset of this expansion of government subsidy of biomedical research have witnessed the introduction of stunning new innovations in health care that have extended longevity and improved our quality of life. Mortality due to cardiovascular disease, while still the principal lethal disorder, has decreased by 60% from its peak in the 1950s. Cancer deaths have declined, and extreme pain and immobility formerly inflicted by chronic joint syndromes are rare. More and more therapies are emerging to treat the "experiments of nature."

Some analysts, however, including myself, have questioned whether this heavily government-subsidized system (which I have named "The Government-Academic-Biomedical Complex—GABC") does not actually inhibit innovation [4,5]. For example, the GABC assumes credit for the emergence in the 1970s of the genetic engineering industry that delivered potent and beneficial biologic drugs, such as erythropoietin, the interferons, and clot-dissolving fibrinolytic agents. But when, after two decades of the GABC existence, relatively little practical innovation was arising from it, Congress enacted legislation designed to encourage patenting of the research discoveries of the GABC grantees and their licensing to startup companies—the Bayh-Dole Act of 1980. Although Bayh-Dole did catalyze innovation by reassuring industry that the government would not appropriate the profits of the licensees, the principal engine behind the biotechnology revolution was insights into genetics, many of which predated the GABC.

One reason the GABC does not obligatorily promote innovation is economic. The largesse of the GABC initially provided and generated an entitled attitude that afflicted researchers and the administrators of their institutions with the fantasy that the party would last forever. The inflationary nature of high-tech research budgets and competing federal priorities conspired to produce a widening gap between supply of and demand for research funding. The gap has resulted in a progressively aging workforce of the GABC-funded academic researchers forced to compete for survival by conforming to the prejudices and risk aversion of grant review committees—hardly a formula for innovation.

Other deficiencies of the GABC are cultural. One that shapes the aforementioned deliberative prejudices is that the problem-solving emphasis of the pre-GABC era of innovation represented a deviation from the behavioral norms that have traditionally characterized science. Historians and sociologists have documented that the principal motivation of scientists—in contrast to technologists—has been to impress influential peers with "novelty," making discoveries of previously unknown entities, and what has been designated "puzzle solving." Puzzle solving includes creating theories that bring natural phenomena into an order that allows for quantitative predictions. The fact that organized science lavishes its awards, such as academic promotion and prizes, on persons accomplishing such achievements is an evidence for the primacy of these motives.

Moreover, medically trained researchers populated the pre- and early GABC workforce and were therefore, intimately familiar with many of the unmet health care needs. But contemporaneous with the establishment of the GABC, riveting basic science discoveries, such as the biochemical basis of genetics, the elucidation of cellular pathways of signaling, and deep insights into subcellular structures, shifted the research emphasis away from more descriptive physiologic and pathologic topics pursued previously to these reductionist subjects.

Although schools teach the rudiments of these reductionist basic science subjects to health care professionals in training, the education is insufficient to prepare them for hands-on experimental work. Adequate preparation has therefore, required students to take time off from more practical training or else—more commonly—defaulted to basic science PhD degree programs in which students without health care experience do science coursework followed by several years of mentored research work. This development conspired with the piling on of ever more lengthy requirements for certification in the subspecialty practices to render combining clinical and research training increasingly unattractive. As a result, individuals with basic science but not medical training have come to dominate the biomedical research population.

Another perverse influence affecting health care innovation is the piling on of burdensome, unjustified regulations affecting the ethics of human and animal experimentation and the financing, conduct, and reporting of research. These rules, however well intended, inflict expense to support the bureaucracies that manage them and cause inordinate time delays. This heavy presence of government oversight due to the dominant federal bankrolling of academic research has been used to justify these inflexible regulations. Performing an experiment as simple as taking a blood sample from a patient and measuring something to test the plausibility of an idea concerning that individual's condition—a frequent procedure in the pre-GABC era—is no longer possible. All too often, by the time an ethics review board sanctions the experiment, the patient is long gone.

These topical emphases and workforce shifts are not categorically counterproductive and have contributed to progress. But capitalizing on basic science discoveries to produce practical innovation requires skill sets and pursuits—described in this book—not taught or rewarded in academic institutions.

Of the long list of underappreciated subjects covered in this book, one that I mention here is the economics of innovation, the lynchpin of which is unlocking investment. Most academics do not even comprehend in the most abstract form the difficulty and high failure rate of innovation. Attracting investment is extremely daunting, yet most academic researchers do not appreciate the particulars of how evaluating market potential and strategies to diminish risk are obligatory activities for obtaining such investment. When a few highly successful biotechnology products licensed by universities delivered large royalties, most academic health care institutions created patenting and licensing bureaucracies. Economic illiteracy and lack of business acumen of both the academic researchers and the institutional offices designed originally to help the investigators explains why inadequately funded and staffed "technology transfer" offices of these institutions rarely succeeded in bringing their bright ideas to practical fruition.

Moreover, as illustrated by the stories of Whipple, Minot, and Murphy recounted above and others, a deep-seated academic belief that maintains that it is only the elusive, complete comprehension of underlying mechanisms that leads to health care benefits is simply false. The tyranny of this belief, however, is that potentially useful inventions may not advance, because the investigators and the institutional offices of technology transfer lack the time consuming and expensive choreography required to define such mechanisms.

The GABC hardly possesses a monopoly on behaviors and misaligned incentives counterproductive to health care innovation. The aspiring innovator must work with private industry to achieve innovation, but this investigator also needs to understand the difficulties that beset that sector. The extraordinary risks referred to above have divided it into two camps. One camp consists of the behemoths that have merged to try and maintain a sufficiently large pipeline to eke out some success despite the high failure rate of drug development. The size of these leviathans guarantees that they are Balkanized, poorly responsive to the need for change, and populated with risk-averse employees. They consistently follow competitors' successes rather than break new ground.

The other camp is the many small- to medium-sized companies developing early technologies. They are more energetic, nimble, and accepting of risk than the battleship industries that these smaller companies hope will acquire them or their technologies, and they probably represent the most exciting homes for the more aspiring health care entrepreneurs. Because entrepreneurship is all about taking risk, the aspirant is likely to live in several such homes before succeeding.

To conclude this introductory chapter, I have emphasized this history, because the impact of what we do today on future events is speculation. The past reveals what has and has not worked to promote meaningful innovation in health care.

Based on the lessons of history, I begin my conclusion by summarizing what I am *not* advocating. I am not recommending that every health care professional engage in innovation. Skilled and compassionate delivery of health care within the confines of today's technologies is an extremely important and noble endeavor—and one all too often not well performed. Researchers with the interests and aptitudes that conform to the traditional status-seeking motives of scientists summarized above should be encouraged to pursue research for research's sake. Economic realities suggest the number of such individuals in academic institutions going forward will decline. Because most research discoveries have little or no impact on innovation, such downsizing may not materially impact health care innovation. That said, individuals committed to careers devoted solely to clinical care or basic science are probably not reading this book.

I am also not advocating abolishing the GABC. It may not thrive as its beneficiaries would like it to, but science has always depended on patronage, whether public or private.

Additionally, I do not believe that all professionals interested in innovation have to default to working in industry. Some folks love doing research whether or not it has practical benefits. Such individuals can function in either the academic or the industrial settings. In the latter, they contribute to trouble shooting and validating projects. Others more focused on innovation need to be aware that industry may not accommodate the long timeline necessary to carry many projects from concept to commercialization and to the clinic. Beholden to their investors, companies cannot let too much blue-sky behavior program their extinction. Researchers employed by private industry—no matter how enamored with a particular research program—often have to drop their work. Provided a committed and persistent investigator outside of industry can garner the resources to sustain a project over the long haul, the relative if penurious security of academe may be a preferable venue.

Finally, while for purposes of clarifying reality, I have enumerated many current impediments to health care innovation, they test the determination of innovators—BUT they are surmountable. The innovation context may have become far more complex than in the pre-GABC era, but on the up side, the opportunities for health care innovation are greater than ever as are the granular roles knowledgeable participants can play to affect it. And, the fundamental drivers of innovation—keen clinical observation and the desire to learn from it—have not changed. This introduction and most of this book emphasize drug and device innovation, but many other vistas loom, one prime example being information technology. The book emphasizes taking ones' own inventions to the clinic, but the innovation system needs participants of many skills, ranging from clinical trial specialists to those involved in and understanding the regulatory arena, management of intellectual property, and financial analysts, not to mention advisors to policymakers to help balance health care access and innovation.

As one who whose career has encompassed many aspects of health care: clinical practice, student teaching, and basic, clinical, and health policy research, I can attest that attempting to convert inventions into clinical products that may save lives has been the most challenging—and the most rewarding. I wish I had been able to read this book early in my career.

REFERENCES

[1] Wootton D. Bad medicine. Doctors doing harm since hippocrates. Oxford, New York: Oxford University Press; 2006.

[2] Starr P. The social transformation of American medicine. New York: Harper; 1982.

[3] Swann J. Academic scientists and the pharmaceutical industry. Baltimore: The Johns Hopkins University Press; 1988.

[4] Stossel T. Removing barriers to medical innovation. National Affairs 2017;(Winter):69–82, http://www.nationalaffairs.com/publications/detail/removing-barriers-to-medical-innovation.

[5] Sarewitz D. Saving science. New Atlantis 2016;(49):4–40, http://www.thenewatlantis.com/publications/saving-science.

Chapter 2

The Basics of Business Law for New Businesses☆

Tom Walsh
Partner, Ice Miller LLP, Indianapolis, IN, United States

FORMING THE BUSINESS

One of the first things an entrepreneur must do is form a business entity through which the entrepreneur can undertake activities to advance the commercialization of his or her medical innovation. The legal forms of business entity that are most commonly under consideration by entrepreneurs are limited liability companies (LLCs) and corporations.[1] The choice of legal entity is one of the most important early stage decisions for an entrepreneur.

Limited Liability Company

A LLC is a legal entity separate from its owners who are called "members." A LLC is formed under the laws of a particular state, which may be the state in which the entrepreneur's primary business activities will be conducted or another state where laws pertaining to management and governance of a LLC may be perceived to be favorable to the entrepreneur.

A LLC is formed by submitting Articles of Organization to the agency of the state government that is responsible for chartering of new businesses. If the Articles of Organization meet the requirements of state law, the state government will issue documentation indicating that the LLC is recognized by the state as a legal entity.

☆ This publication is intended for general information purposes only and does not and is not intended to constitute legal advice. The reader must consult with legal counsel to determine how laws or decisions discussed herein apply to the reader's specific circumstances.

1. The discussion of corporations in this article is limited to corporations that elect to be taxed as corporate entities (so-called "C-corporations"). Corporations also can elect to be treated as pass-through entities for tax purposes (so-called "S-corporations"), but for a number of reasons, S-corporations are not used commonly for entrepreneurial companies.

Medical Innovation. https://doi.org/10.1016/B978-0-12-814926-3.00002-4

Many LLCs also choose to adopt an operating agreement. The operating agreement is a contract among the LLC members and between the LLC members and the LLC itself. The combination of the Articles of Organization and the operating agreement sets forth the details about how the LLC will be managed and governed. While the Articles of Organization are a public record, the operating agreement can be held confidential. Therefore, it is normally in the operating agreement where the members address the most important issues concerning management and governance of the LLC, such as, for example, how managers of the LLC will be appointed, the power and authority of the managers, and the rules concerning the transfer of ownership interests in the LLC.

Ownership interests in a LLC typically are called "units," and are analogous to shares in a corporation. Often, there will be single class of "common" units, however, the Articles of Organization and the operating agreement can establish multiple classes of units, each of which can have different rights with respect to participation in the governance of the LLC, and different rights to the economic benefits that accrue from the LLC's business. For example, the Articles of Organization and the operating agreement can establish that only certain classes of units have the right to elect managers of the LLC or to vote on major decisions affecting the LLC. The Articles of Organization and the operating agreement also can establish that certain classes of units have preferred rights with respect to the economic benefits that accrue from the business of the LLC.

Protection from personal liability is a substantial advantage of doing business through a LLC. Generally, a member of a LLC is liable for the debts, obligations, and liabilities of the LLC only to the extent of the member's invested capital, but under most circumstances the member is not otherwise personally responsible for the financial obligations and liabilities of the LLC.

Another advantage of a LLC is the management and governance flexibility afforded to its members. A typical state LLC statute contains few mandatory rules, so many rules concerning management, governance, and other important and relevant matters can be established by the members of the LLC in the Articles of Organization and/or operating agreement.

A final advantage of a LLC is that a LLC is treated as a pass-through entity for purposes of income tax, which means the LLC itself has no income tax liability. Responsibility for payment of income taxes on the earnings of the LLC is passed through to the members of the LLC. Overall, this structure is more tax efficient than the "double taxation" feature of a C-corporation, but it also introduces risk for individual members of this LLC. Members of a LLC must pay tax on the earnings of the LLC whether such earnings are distributed to the members or retained by the LLC. In other words, if the LLC has a profit in a fiscal year and retains the full amount of the profit for working capital purposes, the members of the LLC will still be responsible for paying the taxes attributable to the profit even though the LLC did not make any cash distributions to the members. The members will be required to satisfy the tax obligation out of other sources of cash. One mechanism that is often employed to avoid this result is to require the LLC to make minimum "tax distributions" to the members. The operating agreement can include provisions requiring the LLC to make distributions to the members at a time prior to the date that the tax returns of the members are due, and in an amount estimated to cover the tax liability of each individual's LLC member.

Corporation

A corporation also is a separate legal entity from its owners who are called "shareholders." Like LLCs, corporations are creations of state law. An entrepreneur may choose to incorporate in the state in which the entrepreneur's primary business activities will be conducted or another state where laws pertaining to management and governance of a corporation may be perceived to be favorable to the entrepreneur. Delaware is a popular state of incorporation, because, among other factors, Delaware's corporate law is well developed (and therefore predictable) and is perceived to be "management friendly" and to promote efficiency in corporate management and operations.

A corporation is formed by submitting the Articles of Incorporation to the agency of state government that is responsible for chartering of new businesses. If the Articles of Incorporation meet the requirements of state law, the state government will issue documentation indicating that the corporation is recognized by the state as a legal entity. Most states also require a corporation to adopt formal bylaws. The rules for management and governance of a corporation are set forth in the Articles of Incorporation and the bylaws.

Ownership interests in a corporation are called "shares" or "stock." There will be a single class of "common" shares. The Articles of Incorporation and the bylaws, however, can establish multiple classes of shares, each of which can have different rights with respect to participation in the governance of the corporation and different rights to the economic benefits that accrue from the business of the corporation. For example, the Articles of Incorporation and the bylaws can establish that only certain classes of shares have the right to elect directors of the corporation or to vote on certain major decisions affecting the corporation. The Articles of Incorporation and the bylaws also can establish that certain classes of shares have preferred rights with respect to the economic benefits that accrue from the business of the corporation.

As with members of a LLC, corporate shareholders will not be liable personally for the debts, obligations, and liabilities of the corporation. The liability of a shareholder is limited to his or her capital investment in the corporation.

When compared to LLCs, shareholders of corporations have considerably less flexibility in designing or influencing how the corporation is managed and governed. The legal rules that regulate the incorporation, operation, management, and dissolution of corporations are the same regardless of the size of the corporation and in many ways are designed for large corporations with numerous shareholders. The result is that for small, closely held businesses, the burdens of a corporation are more complex and involved than those of a LLC. For example, in addition to shareholders, a corporation must appoint directors and officers to manage the operation of the corporation. Directors, officers, and shareholders of a corporation must pay particular attention to observing the "corporate formalities" in order to preserve the limited liability characteristics of a corporation. Observing the corporate formalities requires constant attention to the details of operating a corporation. This important requirement increases the maintenance costs of a corporation.

Management of the business and affairs of a corporation rests with the Board of Directors. The Board of Directors of a corporation is elected by a plurality of the shareholders of the corporation unless the Articles of Incorporation provide otherwise. Typically, the bylaws of a corporation establish the number of directors and the manner in which directors can be elected and removed. The bylaws of the corporation also normally set forth the duties and responsibilities of corporate officers.

Unlike a LLC, a corporation is treated as a separate legal entity for tax purposes. Therefore, the profits of a corporation are taxed twice—first at the corporate level as corporate profits and then again if the profits are distributed to the shareholders as dividends.

Of all of the choices of legal entities, corporations generally provide the most freedom with respect to the transfer of equity interests. Shareholders may transfer shares of a corporation freely to third parties (subject to compliance with securities laws); it however, is not uncommon in closely held corporations for shareholders to choose to contractually limit their ability to transfer shares. A shareholders' agreement is often the tool used to prevent a shareholder from selling shares of the corporation to a third party without the approval of the remaining shareholders (or offering the shares first to the remaining shareholders).

CAPITALIZING THE BUSINESS

After the business entity has been formed and the rights to the underlying technology have been secured, the entrepreneur is faced with what often turns out to be the most challenging part of a new venture—the need to finance the operation and grow the business. In rare instances, a company can be funded solely by its founders without assistance from external sources of capital. Doing so provides founders with the greatest amount of flexibility in terms of running the business, but also requires the founders to take on additional financial risk, because the founders often will need to make substantial financial investments before the company is profitable. For this reason, most entrepreneurs turn to external sources of capital for a new company.

Early Stage Financing

Early stage financing typically involves capital invested to fund the investigation of a market opportunity, the development of the initial version of a product or service, or the pursuit of regulatory approval if the product is a drug or medical device. Typical structures of early stage financing include traditional note financing, convertible note financing, sale of common stock, and sale of preferred stock.

In a traditional note financing, an investor will receive an interest bearing promissory note in exchange for the investment. The interest bearing promissory note normally will contain repayment terms. A note financing is a simple investment to understand, but it is not used commonly in financing new companies for several reasons. Investors normally are seeking a greater return on their investment than the interest payable on a traditional note. In addition, any new company with a limited business history presents a substantial risk of defaulting on the note and likely has few other assets that can serve as collateral for the debt. For these reasons, early stage investors tend to prefer securities that have a greater potential return on their investment.

Convertible note financings are much more common than traditional note financings for early stage companies. Like a traditional note financing, in a convertible note financing, an investor will receive an interest bearing promissory note in exchange for the investment. The convertible note typically will include repayment terms like a traditional note, but the convertible note also will include an option or perhaps a requirement that the amount of the note including accrued interest will be converted into equity (sometimes at a discounted rate or subject to some maximum value) at the time of an equity financing. In other words, investors lend money to a new company, and then rather than getting their money back with interest, the investors receive shares of stock, typically preferred stock, as part of a future round of financing based on the terms of the note.

In addition to the ability to provide investors with the potential for substantial investment returns, convertible note financings have several other advantages that make them attractive for financing early stage companies. The most commonly cited advantage is that a convertible note financing defers discussions of company valuation until the time the note converts into equity. While an investor would not invest without some expectation that the early stage company was going to be successful, it is often difficult for the founders and the investor to agree on a valuation for the early stage company. Without an agreed valuation, it is impossible for the founders and the investor to agree on the percentage of the equity that the investor would receive in exchange for his or her investment. By employing a convertible note, however, the founders and the investor can defer the valuation discussion until a later time when much more information about the company, its products, and its market potential should be available.

Common stock financings are relatively simple to understand. Investors receive a percentage ownership stake in the corporation in the form of common stock in exchange for their investment in the company; however, many investors are interested only in preferred stock in exchange for their investment in the company.

There is no standard definition of preferred stock. Preferred stock generally means that the stock is issued from a class of stock that is different from the "common" class, and that the stock has certain economic rights or other rights that are superior to the common stock. For example, the preferred stock can have a liquidation preference, meaning that the owners of the preferred stock receive distributions from the corporation ahead of the owners of the common stock. In another example, preferred stock can have an antidilution preference, meaning the percentage ownership allocated to the owners of the preferred stock will not decrease even if new stock is issued to other shareholders. In another example, the owners of a class of preferred stock can have a right to control certain governance actions or to appoint a director to the Board of Directors.

The extent to which a company will be required to grant preferred rights to an investor will depend largely on the relative leverage of the parties and on the preferred rights granted by the company in prior financing rounds. It is important to negotiate on the preferred rights that will be granted to an investor. Each financing round creates a precedent. An investor in a later financing round will expect to get all of the rights granted to a prior investor and likely expect to get a better deal.

Grant Funding

While the issuance of debt or equity securities certainly is the most common way to fund a new company, an entrepreneur should not overlook opportunities to fund the new company though nondilutive sources, such as grants from the government and private foundations.

If the technology that the entrepreneur is seeking to commercialize is derived from research funded by the US government, as often is the case with technology that is developed at a research university or nonprofit research institute, then two potential sources of nondilutive funding for the entrepreneurial company are the Small Business Innovation Research (SBIR) and Small Business Technology Transfer (STTR) programs. The goals of the SBIR and STTR programs include stimulating technologic innovation, increasing private-sector commercialization of innovations derived from the US government research and development funding, and fostering technology transfer through cooperative research and development between small businesses and research institutions. The SBIR and STTR programs require a competitive application process and have limited funds. Therefore, an entrepreneurial company seeking financial support through the SBIR and/or STTR programs may benefit from consulting with a party that has experience in preparing SBIR and/or STTR grant applications.

In addition to the SBIR and STTR programs, there also may be grant funding available from private foundations. Such grants from a private foundation, however, are available normally only where the charitable purpose of the private foundation is consistent with the commercial objective of the new company. For example, if the new company is developing a treatment for a certain medical condition, there may be private foundations that exist to fund research and development of treatments for that condition, and those private foundations could be a source for nondilutive funding.

OPERATING THE BUSINESS

The company has been formed and funded, or at least is in the process of acquiring funding. Attention now must be paid to mitigation of risk of the company in its business dealings and development of new assets, particularly intellectual property assets, to advance the company into the future.

Business Contracts

As a company conducts business, it will have relationships with many parties—customers, suppliers, consultants, and employees, to name but a few. Perhaps with a view toward saving money, many new companies enter into contracts with these parties without devoting the same level of time and attention that was given to the formation and capitalization activities. A company's business prospects, however, can be derailed just as easily by a bad contract with a customer, supplier, consultant, or employee, as by a bad operating agreement or investment agreement.

With customer and supplier contracts, appropriate allocation of risk is critical. The rights and obligations of each party must be defined as precisely as possible with consideration of foreseeable contingencies. Risk also is allocated through the structure of the warranty, warranty disclaimer, indemnification, and liability limitation clauses. While many new companies treat these clauses as only legal boilerplate, the language of these clauses will be essential in any dispute that may arise down the road.

Allocation of risk in a contract with a consultant relies on many of the same features—precise definition of rights and obligations, warranty, warranty disclaimer, indemnification, and liability limitation clauses. In addition, clauses of confidentiality and intellectual property rights are important. If a consultant is engaged to work on a nonpublic project, a strong confidentiality clause is necessary to preserve the strategic advantage of the company and also to maximize the rights of intellectual property arising from the project, which may be sacrificed by early publication. Moreover, because intellectual property rights in the United States generally are owned by the party that creates them and must be transferred by written agreement, if a consultant is engaged in work that may result in the creation of new intellectual property, then it is essential to include an assignment of those rights from the consultant in the contract.

Contracts with employees also must cover confidentiality and assignments of intellectual property rights for the same reasons. Unauthorized publication of new ideas or strategic information can be detrimental to the value of the company, and the company must secure the rights to intellectual property created by its employees. In addition, a company may want to consider adding noncompetition and nonsolicitation clauses to contracts with certain of its employees. Such clauses serve to mitigate the risk of an employee using business information or business contacts he or she gained for the benefit of a competitor. One needs to be aware that enforceability of noncompetition and nonsolicitation clauses varies from state to state, so consultation with an attorney who is knowledgeable in this area is critical. In some cases, an invalid, noncompetition or nonsolicitation clause can invalidate the entire contract with the employee, including the critical confidentiality and intellectual property rights clauses.

Enhancing Patent Assets

For many technology companies, developing a strategy to enhance the company's patent assets is an important component of a comprehensive business plan. An effective patent strategy follows four steps. First, the key business goals of the company are identified. Second, the company's existing intellectual property assets are evaluated. Third, new intellectual property is developed. Fourth, the company's intellectual property assets are deployed.

An effective patent strategy begins by identifying the key business goals of the company. Clear business goals provide a long-term blueprint to guide the development of a valuable patent portfolio. In particular, the company should identify the markets in which the company will compete, identify key industry players (competitors, strategic partners, suppliers, and customers), identify technology and/or product directions (within the company and within the industry), determine whether a patent portfolio is to be used offensively (i.e., as a "sword" asserted against competitors or others for licensing opportunities), defensively (i.e., used as a "shield" or for counterclaims against competitors), for marketing purposes (i.e., to show the outside world a portfolio to demonstrate company innovation), or a combination of these, and then meet with a qualified attorney to align goals, industry information, technology/product information, and patent portfolio use.

After the goals of the company are identified, the evaluation process begins by mining and analyzing existing intellectual property assets within the company. For example, for each business goal, the company can determine the core technologies and/or intellectual property that will help to drive that goal. Note that mining and analyzing intellectual property assets involve working with key employees and consultants who can provide input to help align the patent strategy with the business objectives. Here, the company should identify the company's existing intellectual property assets, identify employees and consultants who will create new intellectual property assets for the company, identify the markets and products/product lines for each intellectual property asset, identify those intellectual property assets best suited for patent protection, and prepare a budget for patent strategy and patent procurement by involving a qualified attorney to obtain insights on various costs and fees associated with this step.

While the evaluation phase is in progress, the company can move into the phase of intellectual property procurement. In the procurement phase of the intellectual property strategy, a company builds its intellectual property portfolio to protect core technologies, processes, and business practices uncovered during the evaluation phase. Typically, a patent portfolio is built with a combination of core patents, fence patents, design-around patents, and portfolio-enhancing patents. Each patent may have a unique value proposition for the company. An integral part of the procurement phase is to develop and establish a process for patent procurement and management. An established process improves the ability of a company to evaluate options of intellectual property protection for each innovation, including coverage through patent, trade secret, and copyright. In addition, an established process can be an important component of maintaining cost control.

In the deployment phase, a company should allocate time, money, and resources to further enhance and monetize its intellectual property portfolio. The deployment phase may include licensing all or part of a patent portfolio to others in the industry. Alternatively, it may include asserting rights established through its patents, such as through litigation.

CONCLUSION

The above outline provides one approach to a strategy for a new company pursing commercialization of a medical innovation. As with any strategy, the approach each company may take can differ but should be flexible enough to account for those differences. Every new company, however, will be well served by surrounding itself with competent and experienced advisors, including attorneys, accountants, and business mentors, who can aid in development and execution of the company's strategy and help to mitigate risks along the way. Companies that do these things will be well positioned to capitalize on the rewards for the time, money, and effort spent early on as their business continues to grow and prosper.

IMPORTANT POINTS

- One of the first things an entrepreneur must do is form a business entity—either a limited liability company (LLC) or a corporation.
- An LLC is a legal entity separate from its owners who are called "members." Many LLCs choose to adopt an operating agreement which details about how the LLC will be managed and governed.
- Ownership interests in a LLC are called "units," analogous to shares in a corporation; however, the operating agreement can establish multiple classes of units, with different rights concerning participation in governance and different rights to the economic benefits of the LLC.
- Selected advantages of an LLC include some degree of protection from personal liability and governance flexibility afforded to its members. Also, an LLC is treated as a pass-through entity for purposes of income tax, meaning the LLC itself has no income tax liability.
- A corporation is a separate legal entity from its owners who are called "shareholders" and is formed by submitting the Articles of Incorporation, usually including a set of formal bylaws, to the state governmental agency responsible for chartering of new businesses.
- Ownership interests in a corporation are called "shares" or "stock" with a single class of "common shares," but the bylaws can establish multiple classes of shares, which accrue different rights of participation in the governance of the corporation and different rights to the economic benefits.
- Management of the business and affairs of a corporation rests with the Board of Directors.
- Unlike an LLC, a corporation is a separate legal entity for tax purposes; the profits are taxed twice—first at the corporate level as corporate profits and again if the profits are distributed to the shareholders as dividends.
- One advantage is that corporations provide the most freedom with respect to the transfer of equity interests. Shareholders may transfer shares of a corporation freely to third parties (subject to compliance with *securities laws*).

Chapter 3

Regulating Medical Devices in the United States

David E. Chadwick
Cook Medical, Bloomington, IN, United States

It seems a daunting task, this process of innovation and invention, seeing problems and developing solutions, and putting your sweat and tears and hopes for the future into an idea that you know deep down in your heart will benefit society. You have labored over the idea; you have arrived at what you and maybe some of your most trusted colleagues believe is needed desperately to solve a problem. Your device can make people whole again. Your device can ease the suffering that a patient feels. Your device can possibly decrease the time that one needs to stay in the hospital, off the job, off their feet, or possibly even save their feet from amputation. You have struggled to make prototypes and maybe done bench studies and some animal studies. You believe the new medical device that you have created or to which you have contributed to in its development is ready. Ready to be tested in humans. Or ready to be made available to the US marketplace. Your excitement grows and grows and grows. Something you have conceived may solve a medical problem or prevent one from getting worse. Something you have conceived may be put to use to help a wounded warrior or help a lonely senior citizen live out their final days in a more dignified manner or in a place more to their liking, or your device may ease the pain or help mend broken bones of the youngster who crashed their bicycle into that tree. Your device, your solution, your creation, your innovation.

Then, you meet the US Food and Drug Administration, the FDA—the gatekeepers to the US market when it comes to medical devices. You do not actually meet anyone, but you confront the Regulations and learn about the laws behind the regulations. You learn that there are hurdles to leap over. These hurdles feel like the end of the road for your device, your solution, your creation, your innovation. But fear not. The law does loom large over you, it appears to be insurmountable, and it appears as the very large elephant in the room, that thing, that Government body that now may be blocking you from bringing your solution to help the masses. It looms large in your path. And it is crushing your dreams. But remember, this pathway is important and serves a purpose, and you CAN do it.

Medical Innovation. https://doi.org/10.1016/B978-0-12-814926-3.00003-6

The US FDA is responsible for ensuring the safety and effectiveness of medical products, both drugs and medical devices. Within the FDA, medical devices are controlled by CDRH, the Center for Devices and Radiological Health, which is responsible for "assuring that patients and providers have timely and continued access to safe, effective, and high-quality medical devices."[1] You have learned that the world of interaction with regulators like the FDA, the world we refer to as regulatory affairs, is not so different than your world of medicine or science or innovation. There are consultants and pundits who want to help you.

Sometimes we need to step back and see the FDA regulation as an inevitable and often underestimated challenge. For example, consider the FDA CDRH and how they assist in product innovation. CDRH Innovation seeks to foster the development of medical devices to respond to the unmet needs in public health and to address the new scientific and regulatory challenges of bringing devices to market. The FDA does this important service by increasing the regulatory science capacity of their reviewers and providing assistance to early innovators and small businesses. The FDA is committed to advancing public health by helping to bring innovative technologies to market and providing reasonable assurance that the medical devices already on the market continue to be safe and effective.[2]

Despite this, the CDRH is accused repeatedly by inventors and innovators of inhibiting medical innovation, while at the same time being criticized by others, particularly special interest groups, as not being sufficiently rigorous in reviewing applications for marketing authorization. Opposing views such as these certainly highlight the tension that exists between the benefits of adopting new technology quickly and the risks of unknown and unforeseen safety and effectiveness issues for novel devices. This chapter presents information and strategies to assist inventors and innovators to develop devices by utilizing programs offered by the CDRH. Among these programs are the de novo classification process, presubmission meetings, and early feasibility clinical studies that are designed to facilitate early patient access to innovative medical devices and accelerate marketing authorization of those devices.

BUT, FIRST THINGS FIRST

For me, the first thing is to understand if your invention or new idea is a true medical device or not. Take the smartphone. Is this a medical device or merely a very fancy telephone? In medical device and regulatory jargon, the only answer is "it depends." This response seems to communicate nothing of real

1. https://www.fda.gov/MedicalDevices/ResourcesforYou/Consumers/default.htm.
2. https://www.fda.gov/aboutfda/centersoffices/officeofmedicalproductsandtobacco/cdrh/cdrhinnovation/default.htm.

value and may indeed seem confusing. This response, however, is filled with insight and experience. To help us understand this idea, my preference is to start with and fully comprehend the definition of a medical device. How can one determine the regulatory steps needed to get to market if one is unsure of the rules and requirements of getting to market, especially in our current, highly competitive market of medical diagnosis, therapy, and treatment?

The first step is to determine if your product meets the official definition of a device as presented in the law. If it does, there are FDA requirements that apply. To decide if your product meets this definition, you must understand the official definition.

THE FDA'S DEFINITION OF A MEDICAL DEVICE

Medical devices range from simple tongue depressors and bedpans to complex programmable pacemakers with microchip technology and laser-based surgical devices. In addition, medical devices include in vitro diagnostic products, such as general purpose lab equipment, reagents, and test kits, which may include the technology of monoclonal antibodies. Certain electronic radiation-emitting products with medical application and claims meet the definition of a medical device. Examples include diagnostic ultrasonography products, X-ray machines, and medical lasers. If a product is labeled, promoted, or used in a manner that meets the following definition in section 201(h) of the Federal Food Drug & Cosmetic (FD&C) Act, it will be regulated by the FDA as a "medical device" and is subject to premarketing and postmarketing regulatory controls. According to the law, a medical device is

- an instrument, apparatus, implement, machine, contrivance, implant, in vitro reagent, or other similar or related article, including a component part, or accessory which is:
  - recognized in the official National Formulary, or the United States Pharmacopoeia, or any supplement to them,
  - intended for use in the diagnosis of disease or other conditions, or in the cure, mitigation, treatment, or prevention of disease, in man or other animals, or
  - intended to affect the structure or any function of the body of man or other animals, and which does not achieve its primary intended purposes through chemical action within or on the body of man or other animals and which is not dependent upon being metabolized for the achievement of any of its primary intended purposes.

This definition provides a clear distinction between a medical device and other FDA-regulated products such as drugs. If the primary intended use of the product is achieved through chemical action or by being metabolized by the body, the product is usually considered to be a drug. The FDA has several "centers."

Human drugs are regulated by the Center for Drug Evaluation and Research (CDER) of the FDA. Biologic products, which include blood, blood products, and blood banking equipment, are regulated by the Center for Biologics Evaluation and Research (CBER) of the FDA. The Center for Veterinary Medicine (CVM) of the FDA regulates animal products. If your product is not a medical device but is regulated by another Center in the FDA, each component of the FDA has an office to assist with questions about the products they regulate.[3]

INTENDED USE

Back to the smart phone. Now that you understand the definition of a medical device, the next question deals with YOUR intended use of the device. Is your smartphone merely a fancy telephone or does it capture medical information? As an example, if you have diabetes, and you prick your finger to assess your blood glucose levels and then send your blood glucose readings to your smartphone, your phone is not doing anything with respect to these blood glucose readings to fit the definition of a medical device. But if you set up your phone to transmit these readings to your physician who adjusts your medical therapy based on the readings, you may have elevated your smartphone to the status of medical device. If your smartphone sends data to your external insulin pump (a medical device on its own) and the received blood glucose meter readings change your level or timings of insulin delivery, you have turned your smartphone into a medical device and made it subject to the FDA regulations. Yes, this can be confusing, but as stated earlier, "it depends."

So, what really is "intended use?" What does it have to do with you and your invention? Intended use is a term used in FDA regulation and in the law to express the general functional use of the device, that is, the principal effect of the device's interaction with the body. Alternately, indications for use refers to the specific surgical, therapeutic, or diagnostic use, or group of similar uses of the device, that is, the disease, condition, or pathology for which the principal effect of the device is used to prevent, treat, cure, mitigate, or diagnose.

Your intended use expresses the intent of the product owner or inventor/innovator with respect to their medical device. How do you intend for your device to be used? How do you plan to market or promote your device? Furthermore, your intended use is central to your marketing application to the FDA to gain access to the US market.

PATHWAYS TO MARKET IN THE UNITED STATES

The Medical Device Amendments (MDAs) of 1976 amended the Federal Food, Drug, and Cosmetic Act and provided the FDA with the necessary regulatory

3. See Section 513(a)(1)(B) of the FD&C Act.

authority to ensure the safety and effectiveness of medical devices, as well as establishing the premarket and postmarket regulation of devices.

First, let's examine pathways to market in the United States. The path to market is based on the risk of the device and the perceived controls needed to understand and minimize that risk to patients and users of a medical device. Federal law (Federal Food, Drug, and Cosmetic Act, section 513) established the risk-based classification system for medical devices. In the US, each device is assigned to one of three regulatory classes, Class I, Class II, or Class III, based on the level of control necessary to provide reasonable assurance of its safety and effectiveness.

There are a number of regulatory pathways available to bring medical devices to the market, as published by a group of CDRH scientists (Fig. 3.1) [1] from the Division of Neurological and Physical Medicine Devices/CDRH. The specific pathway to be followed to gain access to the US market is determined by the amount of oversight needed in order to adequately control the risks of a device. Most Class I devices (classed as low risk) require only general controls (Food and Drug Administration, 2014[4]) are exempt from the necessity to

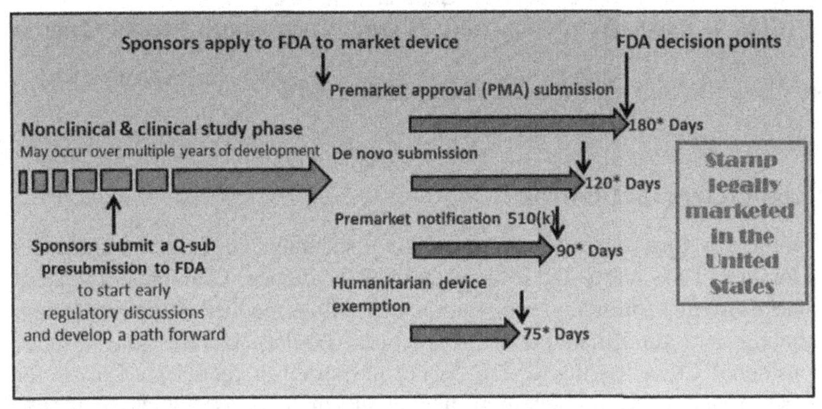

FIG. 3.1 Medical device regulatory pathways to market. An illustration of different regulatory pathways to market, including the investigational device exemption (IDE) phase, where clinical data are collected in human studies. Four marketing submission types are shown, including PMA, de novo, 510(k), and humanitarian device exemption. *The number of days noted is the number of days the submission is under review by the FDA (called "FDA Days"), not the total time that it may take to get the device technology to market or through the review process ("Total Days" is the sum of the time FDA has your submission + days that you are controlling and answering FDA questions). In some cases, the review process may take longer depending on the particular device, technology, indication for use, user, and risk of the device. *(Reproduced with permission from Anderson L, et al. FDA regulation of neurological and physical medicine devices: access to safe and effective neurotechnologies for all Americans. Neuron 2016;92(5):943–8 (Elsevier editor thesis an Elsevier journal!))*

4. General Controls for Medical Devices, https://www.fda.gov/medicaldevices/deviceregulationandguidance/overview/generalandspecialcontrols/ucm055910.htm.

submit an application to the FDA. Most Class II devices (moderate risk) require premarket notification,[5] also known as a 510(k) submission. The reference to 510(k) refers to that section of the law (Federal Food and Cosmetic Act) that created the premarket submissions for these moderate risk devices. Class II devices cleared through the 510(k) pathway require a comparable, legally marketed, Class II device (i.e., a cleared predicate device) to which a sponsor (you) can demonstrate "substantial equivalence" through a comparison of the intended use and the technologic characteristics of the device. In contrast, most Class III devices (high risk) are subject to the premarket approval (PMA) regulations.[6]

As device class increases from Class I to Class II to Class III, the regulatory controls also increase, with Class I devices subject to the least stringent regulatory control, and Class III devices subject to the most stringent regulatory control. Class I devices account for ~45% of medical devices in the US market, Class II devices account for ~45% of devices, and Class III devices account for the remaining 10% of devices.

The regulatory controls for each device class include the following:

- Class I (low to moderate risk): General Controls.
- Class II (moderate to high risk): General Controls and Special Controls (510k).
- Class III (high risk): General Controls and PMA.

Class I, General Controls

A device is Class I if general controls are sufficient to provide reasonable assurance of the safety and effectiveness of the device. Examples of general controls are the following: registration and listing, medical device reporting, labeling, and good manufacturing practices (GMPs). Devices may also be considered Class I if the device "is not purported or represented to be for a use in supporting or sustaining human life or for a use which is of substantial importance in preventing impairment of human health and does not present a potential unreasonable risk of illness or injury."[3] Most Class I devices are exempt from the requirement to submit a 510(k). Examples of Class I devices include manual surgical instruments, tuning forks, and air-conduction hearing aids.

Class I devices do not typically require submission of any product-specific information or product-specific documentation to the FDA.

---

5. Establishment Registration and Device Listing for Manufacturers and Initial Importers of Devices, Subpart E. Premarket Notification Procedures. Title 21: Code of Federal Regulations, Part 807.
6. Premarket Approval of Medical Devices. Title 21: Code of Federal Regulations, Part 814.

Class II, Special Controls

A Class II device is "a device which cannot be classified as a Class I device, because the general controls by themselves are insufficient to provide reasonable assurance of the safety and effectiveness of the device, and for which there is sufficient information to establish special controls to provide such assurance" [1]. Examples of special controls are performance standards, postmarket surveillance, patient registries, special labeling requirements, and development and dissemination of guidelines. Special controls may also include specific types of performance testing (e.g., biocompatibility, sterility, electromagnetic compatibility, preclinical testing) or labeling, which the FDA may outline in the regulation or a special controls guideline.

Most Class II devices require submission and clearance of a 510(k) prior to marketing. Sponsors are required to submit valid scientific evidence in their 510(k) to demonstrate that the device is as safe and effective as a predicate device (a comparable and somewhat similar device made by a company or entity that is legally on the market in the United States). In order to receive marketing clearance, companies that submit a 510(k) for a device must demonstrate how any specified, special controls have been met. Examples of Class II devices include bronchoscopes, vascular catheters and guide wires, bone conduction hearing aids, and external insulin pumps.

Class II devices require typically the submission of product-specific information or product-specific documentation to the FDA describing its intended use. The content and preferred layout of a 510(k) submission are detailed in the regulations (21 CFR 807).[5]

Class III, Premarket Approval

A Class III device is a device which

1. "cannot be classified as a class I device, because insufficient information exists to determine that general controls are sufficient to provide reasonable assurance of the safety and effectiveness of the device," and
2. "cannot be classified as a class II device, because insufficient information exists to determine that the special controls...would provide reasonable assurance of its safety and effectiveness," and
3. "is purported or represented to be for a use in supporting or sustaining human life or for a use which is of substantial importance in preventing impairment of human health," or
4. "presents a potential unreasonable risk of illness or injury."[6,7]

7. See Section 513(a)(1)(A) of the Food, Drug and Cosmetic (FD&C) Act.

Class III devices require a premarket approval (PMA) application prior to marketing and valid scientific evidence must be provided to demonstrate a reasonable assurance of safety and effectiveness. This evidence must be submitted in the form of a PMA application. Examples of Class III devices include cochlear implants, implantable cardiac defibrillators, and vascular stents.

Class III devices always require submission of product-specific information or product-specific documentation to the FDA. In addition, Class III devices always require the submission of human clinical data obtained to demonstrate the safety and effectiveness of the product for its intended use. The content and preferred layout of a PMA application are detailed in the regulations (21 CFR 814).[6]

Preamendments Devices

Medical devices on the market prior to the Medical Device Amendments of 1976 were allowed to remain on the market if they were deemed safe. The term "preamendments device" refers to devices legally marketed in the United States by a firm before May 28, 1976 (the date that President Gerald Ford signed MDAs of 1976) and which *have not been*

- significantly changed or modified since then
- for which a regulation requiring a PMA application has not been published by the FDA

Devices meeting the above criteria are often referred to as "grandfathered" devices and do not require a 510(k). The device must have the same intended use as that of another device marketed before May 28, 1976. If the device is labeled for a new intended use or if the device is changed substantially or modified, then the device is considered a new device, and a 510(k) must be submitted to FDA for marketing clearance.[8]

ACTIVITIES TO SUPPORT MEDICAL DEVICE INNOVATORS

Innovative medical devices often present new scientific and regulatory challenges for small businesses and start-ups.[9] The CDRH Innovation has a twofold approach to help to increase patient access to innovative medical devices developed by medical device innovators to (1) increase outreach to these innovators and (2) increase training opportunities for regulators to learn about the unique challenges faced by start-ups.

---

8. Preamendment status, https://www.fda.gov/medicaldevices/deviceregulationandguidance/medicaldevicequalityandcompliance/ucm379552.htm.
9. https://www.fda.gov/AboutFDA/CentersOffices/OfficeofMedicalProductsandTobacco/CDRH/CDRHInnovation/ucm516820.htm#early.

The FDA provides resources to encourage medical device innovators to receive input from the CDRH early in the process of device development and to increase training opportunities for CDRH staff to help promote patient access to innovative devices and decrease the costs from concept to commercialization for innovative medical devices.

EARLY REGULATORY ASSISTANCE FOR MEDICAL DEVICE INNOVATORS

The CDRH offers two meeting options to answer questions you may have about the marketing of your specific device

- *Informational meeting*—You may request a meeting to share information with the FDA to provide an overview of ongoing device development or to help familiarize the review team with new devices that have substantial differences in technology from currently available devices.
- *Presubmission program*—You may request formal feedback from the FDA. The feedback may be provided through a face-to-face meeting, teleconference with feedback documented in minutes of the meeting, or in a written response. A presubmission is appropriate when feedback from the FDA on specific questions is necessary to guide product development and/or application preparation.

In addition to asking traditional questions about device design, medical device innovators are encouraged to ask questions about manufacturing and quality systems during their presubmission meeting. The quality system regulation (21 CFR 820) applies to finished device manufacturers who intend to distribute medical devices commercially and covers design, manufacturing, and other quality system activities.

If you plan to conduct research involving human subjects, information about good clinical practices and the protection of human subjects is available at Device Advice: Investigational Device Exemption (IDE).[10] If you have specific questions about the regulatory process for medical devices, contact the Division of Industry and Consumer Education (DICE) or for general regulatory information, visit Device Advice.[11]

REMEMBER YOUR AUDIENCE

A few final thoughts as you move from invention to interacting with FDA. Remember your audience.

10. https://www.fda.gov/MedicalDevices/DeviceRegulationandGuidance/HowtoMarketYourDevice/InvestigationalDeviceExemptionIDE/ucm2005715.htm.
11. https://www.fda.gov/MedicalDevices/DeviceRegulationandGuidance/default.htm.

The reviewers at the FDA do not know your new medical device. If the technology is brand new and something they have never seen before, they will not understand what you are "bringing to them in paper form" as your submission. Your job is to provide enough information and use appropriate language based on what your audience (FDA reviewers and scientists) does and does not know.

Communication may be the most important skill a person can have, and according to John Maxwell, a leading thinker in the world of leadership and communication, focusing on the other person, whether one on one or in a group, is the best way to communicate with them. Maxwell has been quoted[12] on communication to say, "If I had to pick a first rule of communication-the one practice above all others that opens the door to connection would be to look for common ground." In other words, keep your audience in mind when you craft your message.

Your job is to tell the FDA what information they need to know and how they can use this information to make sound decisions on the applicability of your new product to the US market.

In summary, FDA is not really an onerous roadblock in your pathway. They are an agency of the Federal Government tasked with protecting the American public. The laws they follow are passed by the Congress and signed by the President. The subsequent regulations they promulgate are based on these laws and are there to protect you and me and all other Americans. The FDA regulates our foods, drugs, medical devices, and other major areas of our economy with the primary purpose of keeping us safe and guaranteeing that the products they regulate, including the result of your innovation, are safe and effective for its intended purpose. They also stay involved with your product after it enters the US market. They monitor that you and your competitors play on a level playing field and that your promotional activities accurately represent the marketing clearance or approval you and your competitors received. It may appear that the FDA and their accompanying regulations will hinder you, but they are there to assist you in the development and commercialization of your products. They know you have put your sweat and tears and hopes for the future into an idea that you know deep down in your heart will benefit society. They will assist you, you need only ask and read the publications they make available.

IMPORTANT POINTS

- This chapter presents information and strategies to assist inventors and innovators in the development of devices by utilizing programs implemented by the CDRH.
- The US FDA is responsible for ensuring the safety and effectiveness of medical products, both drugs and medical devices. Within the FDA, medical devices are controlled by CDRH, the Center for Devices and Radiological Health.

---

12. http://www.azquotes.com/quote/858285.

- Determine if your product meets the Official Definition of a device as presented in the law.
- Intended use is a term used in FDA regulation and in the law to express the general functional use of the device.
- The specific pathway to market is determined by the degree of oversight needed in order to adequately control the risks of a device.
- Each device is assigned to one of three regulatory classes, Class I, Class II, or Class III, based on the level of control necessary to provide reasonable assurance of its safety and effectiveness.
- The FDA provides resources to encourage medical device innovators to receive input from the CDRH early in the process of device development.
- Remember your audience. Your job is to provide enough information and use appropriate language based on what your audience (FDA reviewers and scientists) does and does not know.

REFERENCES

[1] Anderson L, et al. FDA regulation of neurological and physical medicine devices: access to safe and effective neurotechnologies for all Americans. Neuron 2016;92(5):943–8.

Chapter 4

The Role of University Technology Transfer

Jay Schrankler
University of Minnesota, Minneapolis, MN, United States

A new medical device, biologic, therapeutic, or drug can arise from many places, including practicing medical professionals, individual inventors, and corporations. Perhaps not surprisingly though given their focus on learning and new knowledge, researchers and clinicians at universities have been behind some of the most impactful medical inventions.

There are three aspects of the technology transfer organizations that are key to getting an idea from the lab to the market and into clinics, doctor's offices, and hospitals:

(1) *Functions and people*—the ideal makeup of the staff of these organizations and the functions within.
(2) *Process*—identifying and detailing the process used in technology transfer to get an idea to health care practice settings.
(3) *Programs*—specific programs used to encourage innovation and entrepreneurism among the research and clinical community and to stimulate industrial collaboration with research.

Each of these key aspects will be covered in order.

It important to know how a University technology transfer office functions. The general, "high level" view of how inventions are "sought" from inventors and how they get to a "marketable" state is depicted in Fig. 4.1.

FUNCTIONS AND PEOPLE

Each technology transfer office will have slightly different functions and staff makeup; however, the best performing groups have similar functions and requirements for these functions. An older model in technology transfer is the "cradle-to-grave" functional model, in which a single individual, typically

Medical Innovation. https://doi.org/10.1016/B978-0-12-814926-3.00004-8

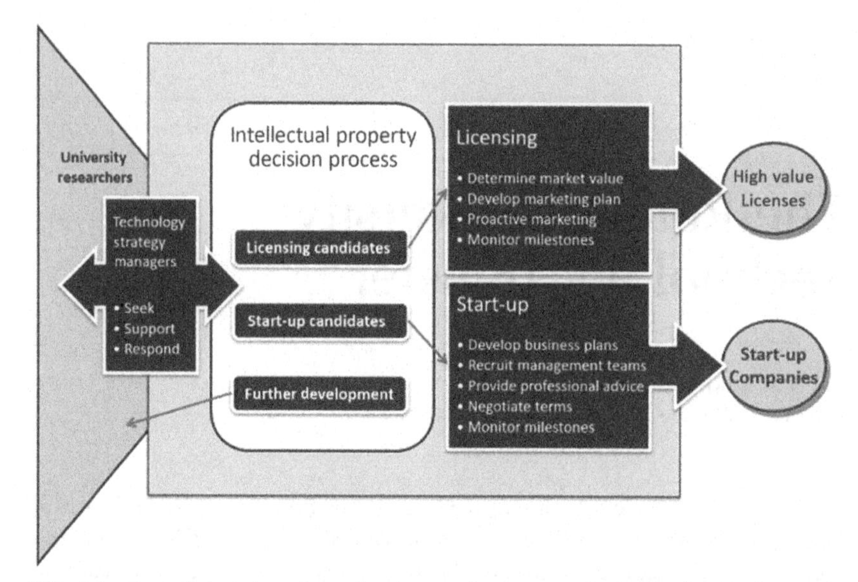

FIG. 4.1 General depiction of a university transfer process—from Idea "seeking" through decision-making to licenses and start-up companies. Sometimes, ideas need more work and are sent back to the researcher for further development.

called a *Licensing Manager*, takes an idea (invention disclosure) and brings it through the entire process to market. Many smaller universities adopt this model out of necessity.

A newer model used more in larger universities is a functional one. The necessary distinct skills and the types of backgrounds needed for these functions are listed below. Different universities will have unique job titles, but the function usually falls into one of these roles:

A Technology Strategy Manager works with University researchers to seek out new inventions, to encourage and support researchers, and to move new invention disclosures through the first stages of commercialization. Successful technology strategy managers have a strong foundation in science or engineering as well as industry experience, often as a product manager in a larger corporation.

An Intellectual Property (IP) Manager handles patent, trademark, copyright, and other aspects of intellectual property of an invention and works hand in hand with a technology strategy manager to move an invention or "case" forward by providing legal support of the IP. In addition, the IP manager handles the regulatory pieces of federally funded research, because patents that arise from such research require substantial and careful compliance reporting to the US Government. Typically, IP managers have a legal background, such as serving as a patent agent, paralegal, or IP attorney.

A Technology Marketing Manager markets inventions to industry. These individuals work closely with technology strategy managers and are handed the case once the IP has been protected legally. The technology marketing manager develops a marketing plan and materials for the invention and also includes an inventor or inventors to refine an invention to meet the needs of identified market niches. Technology marketing managers often have a background in sales or marketing in industry, and most have experience in science or engineering.

An Office Marketing Manager is responsible for "spreading the word" about the portfolio of technologies in the technology transfer office. This individual handles overall marketing of the technology transfer function, with a heavy emphasis on digital/online marketing, as well as internal and external events. A major component of this role is getting descriptions of inventions posted to websites, including their institution's own online roster of available technologies as well as external sites, such as the Association of University Technology Manager's Global Technology Portal and ibridge copyright.

If a technology appears best-suited for a new start-up or spinout company, a *Venture Development Manager* helps assemble a team to launch a new company. The venture development manager works closely with both technology strategy managers and technology marketing managers in order to manage the IP and the license typically associated with a new venture. Most venture development managers are either seasoned entrepreneurs or have been a partner at a venture firm. In the university setting, a fundamental background in science or engineering is highly preferred.

Other functions or titles may occur in technology transfer offices—as shown earlier, all these experts are the key functionaries in technology transfer. Note, smaller universities will have licensing managers that take on many of the above roles. In addition, each university will have its own unique job titles. It is more important to understand the function than to be concerned about the specific title.

In smaller technology transfer offices or in smaller universities where technologies are often handled by a single individual from "cradle to grave," a *Licensing Manager* takes on the role of both technology strategy manager and technology licensing manager, as well as parts of the other functions described earlier; often, however, this single person may struggle with one or more of the varied roles along the commercialization timeline.

If you are a researcher or clinician in a university interested in creating new inventions, it can be helpful to meet in the ideation phase and develop a relationship with the technology commercialization professionals described above, and to work closely with them on new ideas.

PROCESS

Although some aspects vary between institutions, the process of moving a university invention to the market has some standard practices. Many best practices have been identified in more than three decades since the passage

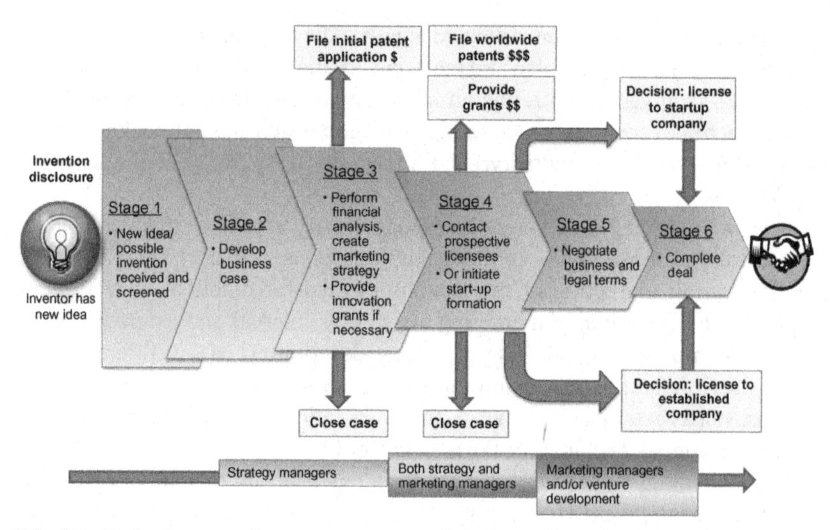

FIG. 4.2 Technology transfer stage gate process. Bringing an invention from idea inception to market.

of the Bayh-Dole Act in 1980. Bayh-Dole gives universities the authority to commercialize inventions resulting from federally funded research that faculty and staff create with specified rights and royalties for the inventor, the university, and the federal government. Fig. 4.2 shows a typical process for an invention.

Fig. 4.2 outlines a "stage-gate" process that is also common in companies and start-ups. Each stage describes a specific part of the commercialization process, and each gate is a key decision-making point—a "go-no go" decision to move on to the next stage.

Stage 1: Invention disclosure (*new idea*)—Inventors need to "disclose" their idea. This an important part of the legal process toward protecting an invention. This disclosure is made typically to the technology transfer office at the university, usually to the technology strategy manager or licensing manager. An inventor in a US university setting *must* disclose the idea to the technology transfer office orally or in writing before any public disclosure. If an idea is not disclosed and subsequently not protected before public disclosure, the protection for the idea is, more often than not, lost forever. Most technology transfer offices have specific forms for filing such discourses and will provide assistance to inventors in this stage of the process. Typically, technology transfer offices will screen ideas at this stage for public disclosure or other regulatory or contractual issues the researcher may have.

Stage 2: Developing the business case—During this phase, the technology transfer office will work to assess the potential business opportunity for the invention. In the case of a medical technology, it may also make an initial assessment of the regulatory path and the possibility of reimbursement for the use of the invention from health insurers. Whether it will be an expensive and lengthy regulatory process, the initial likelihood of reimbursement can make a huge difference in whether an invention is commercially viable. Another key component of this phase is an initial search for issues concerning "prior art" and potential "freedom to operate"; in other words, are there other patents or preexisting published results that may interfere with the protection of the invention? In order to avoid the costs of a full legal assessment early on, this process represents a necessary, albeit cursory, look at existing evidence; these assessments may be performed through simple web searches. At this stage, too, the invention will be assessed for "patentability," examining whether or not it passes the threshold of being patentable subject matter under US law. For the business case assessment, there is often a first assessment of the market for the invention using professional market research tools (most offices will likely start the marketing process at this point or at least gather materials and research for marketing). The invention must have a market size that justifies the expense of protection of the IP, and the commercial investment. Finally, the technology transfer office will usually make a decision on whether or not to file a "provisional patent" application on the invention. A provisional patent application is essentially a place holder in the patent office and will hold the place for the invention for 12 months. It also entails relatively little expense. After the year of the provisional patent, another decision must be made as part of Stage 3, and here the decisions become much more expensive. Within this context, usually technology transfer offices have a defined annual budget for their patent filings, which means they need to consider each filing in the context of an overall portfolio of patents.

Stage 3: Detailed market and financial analysis, potential grants for gap funding, and a key decision to convert the provisional patent to a utility patent or PCT (patent cooperation treaty)—At this stage, there are many moving parts. First, it is typical for the technology transfer office to have done further analysis on the invention and the market. It is important for the inventor to be involved during this part of the process. Important work on marketplace fundamentals as well as a detailed financial analysis of the market occur at this time. This analysis will be used to make key decisions on patent conversion to a Utility Patent or a PCT. Second, the technology transfer office and IP manager do further work, often with legal counsel, on the patent, keeping the inventor "in the loop" on the construction of the patent claim(s) and prior art analysis. Third, for a small percentage of ideas, the office may secure and/or provide gap funding to get more data (helping both the patenting and marketing process) or to further develop the invention. In the medical field, a common use for such funding is to get more animal-based laboratory data or to build a prototype and test it. Not all technology transfer offices have access to gap funding. If it is needed but not available from the technology transfer group, then other sources are sought. Finally, a decision is made typically by the technology transfer office to file a utility conversion of the provisional patent or to file a PCT application. Each of these decisions costs much more than the provisional filing done in Stage 2. The inventor needs to be aware that once a utility patent or PCT is filed, the patent will now progress through prosecution. In prosecuting a patent, an examiner in the patent office will be assigned to the case. The examiner will likely have "office actions" to which the inventor (and technology transfer office) must respond. These are challenges to the claims in the application, and they must be answered to the satisfaction of the examiner for the application to proceed. Each office action has an associated cost. Successfully getting beyond Stage 3 means the invention is ready to be commercialized and hopefully start a path to the market in earnest.

Stage 4: Marketing a license to an existing corporation or a decision for formation of a start-up company—This is the stage where the serious marketing efforts begin, with heavy involvement from the technology marketing managers and also the inventor, who plays a vital role in meetings with potential licensees—that is, companies who are interested in using the technology. Inventors play a critical role in those meetings not only to help get a deal done but to understand the role they, as the inventor, will need to play once a license agreement has been signed. Even more important to the marketing effort for the inventor is the role they can play in generating leads for the technology transfer office. Many inventors have strong connections with industry or have worked on sponsored research projects with industry. These relationships and connections can serve the marketing process well.

Some inventors may decide they would rather travel down the entrepreneurial path and work toward using their invention as the basis for a start-up company. Partnering with the technology transfer office and the venture development manager is a good choice in these situations. Such a partnership(s) can provide substantial insightful support, training, and resources. In addition, the technology transfer office can inform the inventor of the market analysis and whether or not their invention is disruptive and unique enough to become the basis of a start-up company.

Stage 5: Negotiating the business and legal terms of a license agreement—This stage is led by the technology transfer office, typically by the technology marketing manager or the licensing manager and is often assisted by a transaction (or contract) attorney from or designated by the university's general counsel's office. If the license is with an existing company, then it is a good idea for the inventor to stay abreast of contractual terms and conditions during the negotiation and to be an active partner to the technology transfer office. In contrast, if the transaction is leading toward a start-up company where the inventor will have an important and substantial role, then it is usually better that the inventor is recused from the process to avoid a conflict of interest during the negotiation. Another key point regarding start-up companies is that the inventor stands to benefit from equity and royalties associated with the license. If inventors take up a prominent role with the start-up company, they may take equity in that company outside of the license agreement the company has with the university. Many times the inventor will not want to leave their faculty or clinical appointment and will decide to work with the technology transfer office in the search for an experienced entrepreneur to actually run the company. In contrast, the inventor may want to be the "entrepreneur" and will need to commit much more time and effort to be a more integral part of this process.

Stage 6: Completing a deal and transferring the technology and associated knowledge—This is the final stage in the process. The technology transfer group and transfer organization completes the contractual paperwork, which means licenses, agreements, and start-up documents, where needed. It is crucial at this stage for inventors to become familiar with the license agreement, with the assistance of the technology transfer group. Often in these agreements, there are requirements listed for the inventor to fulfill. These might include data and knowledge transfer and requirements related to improvements to the invention in the future.

Finally, there are requirements at most research institutions for conformance to conflict of interest policies. This can be a large factor in what is allowable and not allowable for inventors regarding direct participation in follow on activities with a licensee and/or start-up company. It is a good practice and potentially very important for the inventor and his/her future interests to get involved in the process of COI (conflict of interest) early in the licensing process. Many technology transfer organizations can be quite helpful with inventors in getting them through the COI requirements. Typically, COI is not an issue; however, in the case of clinical trials, it can become very complicated, and, thus, it is a good practice to get involved early in the process.

Waiver of all or part of Intellectual Property Rights, Title, and Interest: Commercialization of new technology is not a certain process that always ends in success. There can be cases where the inventors themselves may be more successful. To address this matter, entities that are charged with the commercialization process will often offer to waive all or partial rights to the inventors. Many of these organizations will do so after they have exhausted all other possibilities, and the inventors have requested a waiver of rights back to them. Owners of the intellectual property, universities, hospitals, etc., will review these requests and often grant waivers. It is important to keep in mind that each of these entities will likely have their own policies regarding waivers. Options of waivers can include all fields of use or only certain fields of use. Inventors need to be aware that there may be costs, such as sunk patent costs, that will need to be covered either by them personally or by other funds that they can acquire. In addition, if there is more than one inventor for the IP being waived, they must decide, as a group, how they will handle management of the invention going forward. Options include forming an LLC, partnership, or assigning it all to one of the inventors. Most organizations can and will only waive to the inventors of the IP.

PROGRAMS

Moving an invention from a research institution to the market is not an easy task. To make this process easier and to help the aspiring researcher-entrepreneur, many organizations offer programmatic assistance.

Innovation and Entrepreneurship

The *lean start-up* and the related lean business model canvas have been adopted as a de-facto standard for developing the framework of a business for an invention. Even more generically, this method can assist inventors in thinking about how their invention can be developed, so that it does get closer to meeting the needs of the marketplace.

An example in the United States is the NSF (National Science Foundation) icorps initiative to expand the expertise of students and scientists in the field of STEM (Science, Technology, Engineering, Mathematics) beyond traditional laboratory skills and to build its participant's experience in translation of discoveries to the commercial world (https://www.nsf.gov/news/special_reports/i-corps/ and https://mincorps.umn.edu/programs). These programs are optimal when interdisciplinary in nature, especially when they bring together research departments and the experts from the business school or industry. Many programs offer a range of classes and coaching sessions in areas such as Lean Innovation principles, value proposition design, and start-up strategy specifically involving the practice of early customer feedback on concepts and ideas.

Key to these programs are the ranges of academic and business experience among the leadership team of the technology transfer office, and the link with advisors and mentors from the business community whose expertise is invaluable in guiding participants as they work to identify commercially relevant applications and markets for the cutting-edge inventions they are developing.

The important take away message is for inventors to seek out a program and/or additional support to assist them with the process of innovation and entrepreneurship. A good place to start is the emerging standard practice of the lean canvas. Even if not part of a formal program, there are many self-directed resources available for this important step.

Another programmatic example specifically for medical innovations is the US NIH (National Institutes of Health) REACH program (https://mn-reach.umn.edu/). The REACH program (Research Evaluation and Commercialization Hubs) provides innovators the opportunity to create connections within and beyond the University, strengthen existing alliances between stakeholders (including public, private, nonprofit, and academic sectors), and create cultural and systemic changes to more rapidly move from breakthrough innovations to products that will have health, economic, and societal impact.

REACH involves a crucial skills-development component in its programmatic materials, from boot camps to half-day workshop series and monthly strategy clinics, all with relevant clinical, regulatory, and reimbursement experts. All these elements combine to assist researchers on how to position their research for maximum commercial relevance in the health care industry, and how to bring their technologies to commercial reality.

At each institution awarded a REACH grant by NIH, the program is supported by a board of highly qualified individuals from within academia and experts from the therapeutic device and biotech industries. This board interfaces with the researchers and assists in the development of funding of translational proposals.

Industry Supported Research and Commercialization

Innovation partnerships with medical industry companies can be a great collaboration and funding opportunity for researchers. Likewise, companies can benefit greatly by collaborating with leading edge researchers. Often, these opportunities can lead to important commercial opportunities that lead to clinical applications.

Traditionally, one key issue in these collaborations has been the rights to resulting IP developed during the collaboration. For this reason, many leading research institutions have developed new models for industry collaborations that make it easier for IP development and any resulting rights. In fact, it is the technology transfer offices at these institutions that have led the way in this important new programmatic development.

As a researcher becomes interested in collaborating with industry, it is very important to work with the technology transfer office in determining resulting IP rights. Understanding this before such a partnership is begun is the best option. Looking at new models for industry partnerships is critical. One good benchmark is the MN-IP program (https://research.umn.edu//units/techcomm/sponsoring-research-mn-ip) developed at the University of Minnesota and replicated at other universities.

Another example of industry support and collaboration is the Earl E. Bakken Medical Devices Center (Bakken MDC) at the University of Minnesota. The Center performs basic and applied research related to medical devices, and trains a cohort of Medical Device Innovation Fellows each year to be leaders in the field. The fellows are taught the disciplined process of product development needed for medical devices, including FDA requirements, insurance reimbursement, IP, and business strategies, as well as creativity techniques and prototyping. The Innovation Fellows have been very successful, generating over 200 invention disclosures within the first 8 years, and they, along with the leadership of the center, have exposure to the technology transfer office, operating rooms, and development labs around the University (http://www.mdc.umn.edu/).

CONCLUSION

In today's world, technology transfer offices are highly valued members of the innovation community within universities. Their work is often technically demanding, and to understand the technology in a potential new idea, they must also be keenly aware of the market for that technology. In this field, we have learned a great deal through trial and error over the years, and have developed processes, functions, and programs to help nudge and accelerate ideas from the lab to everyday use. Among universities, we continue to learn from each other's work, and we continue to refine how we do things to make the best use of scarce commercialization resources. Now more than ever, and especially among junior faculty, we see researchers' interest in creating new inventions and start-up companies, and we in technology transfer share their enthusiasm and their interest in a better future.

IMPORTANT POINTS

- Technology transfer offices have two models: a "cradle-to-grave" model where one person, the *Licensing Manager*, shepherds an idea (invention disclosure) through the entire process to market (usually in smaller universities), and a second newer, more functional model where individual specialists work together to push the innovation forward.
- Important specialists in a technology transfer office include
  - **(a)** Technology strategy manager seeks new inventions, supports researchers, and moves the new invention disclosures through the first stages of commercialization.
  - **(b)** Intellectual property (IP) manager, often a lawyer, handles patent, trademark, copyright, and other aspects of IP including the regulatory pieces of federally funded research.
  - **(c)** Technology marketing manager has experience in marketing and science/engineering and markets inventions to industry.
  - **(d)** Office marketing manager responsible for "spreading the word" externally about the portfolio of technologies in the technology transfer office with a heavy emphasis on digital/online marketing and internal and external events.
  - **(e)** Venture development manager assembles the team to launch a new company if the innovation is best suited for a start-up or spinout company.
- The process of moving a university invention to the market is complex and involves
  - **(a)** Invention disclosure.
  - **(b)** A market and financial analysis with potential grants for gap funding, and a key decision to convert the provisional patent to a utility patent or PCT (Patent Cooperation Treaty).

(c) Marketing to license to an existing corporation or a decision for start-up company format.
(d) Negotiating the business and legal terms of a license agreement.
(e) Completing a deal and transferring the technology and associated knowledge.
(f) Conformance to conflict of interest policies.

- Many institutions and national organizations offer programmatic assistance in innovation, entrepreneurship, and protection of IP including
  (a) NSF (National Science Foundation) icorps initiative to expand expertise of students and scientists to build experience in translation of discoveries to the commercial world (https://www.nsf.gov/news/special_reports/i-corps/).
  (b) US NIH (National Institutes of Health) REACH program (https://mn-reach.umn.edu/) which provides opportunity to create connections within and beyond the University.
  (c) MN-IP program (https://research.umn.edu//units/techcomm/sponsoring-research-mn-ip) which helps to suggest new models for industry partnerships.
- Finally the innovator/inventor/prospective entrepreneur at a university MUST recognize that in today's world, they will need the technology transfer office which contains highly valued and experienced members of the innovation community within the university.

ACKNOWLEDGMENTS

I thank Dr. Thomas Hutton from the University of Minnesota Office for Technology Commercialization and Dr. Joseph Hale from the University of Minnesota Medical Devices Center for their contributions.

Chapter 5

Basics of Patent Law: Strategies for Entrepreneurs and Start-Up Enterprises

Thomas J. Filarski
Steptoe & Johnson LLP, Washington, DC, United States

Abraham Lincoln once proclaimed that "[t]he patent system added the fuel of interest to the fire of genius." While the observations of President Lincoln over 150 years ago may be inspirational to innovation, obtaining a patent for a new idea or creating a successful business for a new product or process based on that idea is a challenge.

This chapter will explain the fundamentals of protecting an idea under the United States patent system and will identify pathways to bring that idea to the public. The goal is to help the inventor understand the elements of a patent, what is required to obtain one, and the value of a patent and its use in today's business environment.

WHAT IS A PATENT?

A United States patent is issued by the government through the United States Patent and Trademark Office (USPTO). It has been said that a US patent emerges from a covenant with the government; the inventor discloses an invention and in exchange, the government grants the right to the patent holder to exclude others from using the patented invention for a period of time.

In fact, a US patent gives the holder the right to exclude others from making, selling, or otherwise using the subject matter that is covered by the claims of the patent. The patent laws require the inventor to disclose the invention in a certain manner and to show that the claimed invention is different from what others have done before. If these requirements are not honored, a patent may be revoked by the USPTO or the patent can found to be invalid in a federal court proceeding. A US patent is enforceable for a specific period of time within the borders of the United States and its territories. If another person or company

Medical Innovation. https://doi.org/10.1016/B978-0-12-814926-3.00005-X

uses the patented idea without authorization, the patent owner may sue in federal court for a determination of infringement and seek restitution, including monetary damages and an injunction against the infringer. At the time of its expiration, the subject matter claimed in the US patent enters the public domain, and the exclusivity rights held by the patentee terminate.

A patent has many uses in commerce and is an asset that can support a business. A patent can be bought, sold, licensed, pledged, or used otherwise to obtain funding for the development, manufacture, marketing, and other needs of an enterprise. A patent promotes and aids in branding a technology-focused company, and it supports a greater price for a product, allowing for premium over commodity pricing.

A patent also presents a barrier to competitors of a new product or service introduced by the patent owner into the US marketplace. A US patent encourages a scrupulous competitor to design around or differentiate its own product or service from the invention claimed in the patent. If not, the competitor faces an enforcement proceeding from the patent owner which is intended to stop unauthorized use of the patented invention. A patent also may deter litigation against the patent owner, because patents are often considered to be defensive bargaining chips that can encourage settlement of a dispute with a competitor and lead to cross-licensing of technology.

Importantly, a US patent does not give its owner the right to practice the invention that is claimed in the patent. This clause is important, because a patent gives the holder a right to exclude, but it does not give the right to make or use something. The fundamental concept of exclusivity—that a patent provides the right to exclude others from practicing, but does not guarantee the right to practice the invention that is claimed in the patent—was built into the patent laws purposely and is key to understanding the use and value of patents to an enterprise. The principle of exclusivity, for example, fosters improvements in technology and can lead to licensing of the patent in exchange for revenue or other technology.

Thus, it may occur that a particular product or service that includes the patented invention also contains features that are covered by a different US patent that is held by another entity. In such an instance, the other patent dominates the particular subject area and as a result, precludes others from using a feature that is critical to the technology. Thus, without the authorization of the other dominant patent holder, the particular product will be sold or used at the risk of infringement, which can lead to the cost of a patent litigation and the threat of monetary damages. Such a situation presents an opportunity for a "design around," where products or services are changed substantially or differentiated from, and at times improved over, the patented invention. Alternatively, this situation may lead to a license in which the patent owner grants to the licensee the right to make, use, or sell the product or service covered by the dominating patent in exchange for revenue or other valuable assets.

SECURING PATENT RIGHTS

To obtain a patent, one must describe the idea in an application for examination by the government. The patent examination is normally open to the public, and the applicant must show that the claimed invention is different from what others have done before.

The first step in the patent application process is to retain a patent attorney or patent agent. That person's task is to prepare a written description of the invention and communicate with the USPTO to obtain the patent. Whether an attorney or a patent agent, these professionals have passed an examination overseen by the USPTO and are registered to practice before the USPTO. The USPTO also requires certain technical training to qualify as a registered agent or attorney, and because backgrounds and experience vary widely, the professional should be interviewed and selected to fit the area of the invention and the particular needs of the inventor.

The USPTO allows an inventor to prepare, file, and prosecute his/her own patent application. But the US patent laws and regulations can be complicated, and it is highly advisable to seek professional assistance for this task. A list of registered patent agents and attorneys is available from the USPTO website. https://www.uspto.gov/patent. A search of the Internet will return lists and blogs with reviews and recommendations of patent practitioners.

TYPES OF PATENTS

There are three main types of US patents: utility, design, and plant. A *utility patent* seeks to cover the functional features of an invention that are novel and nonobvious (35 U.S.C. §§ 101–103). As a general rule, a new utility patent application, once allowed, will have a term beginning from the date of grant up to 20 years from the date the first application was filed with the USPTO, known as the priority date. There are exceptions to this rule, and the law provides for opportunities to extend and add time to the term. In contrast, a *design patent* covers any new, original, and ornamental design for an article of manufacture (35 U.S.C. § 171), and a new design patent application will have a term of 15 years from the date of grant. Finally, one may obtain a *patent* for any distinct or new variety of a *plant* (35 U.S.C. § 161), the term of which is similar to that of a utility patent. All of these patents give the patent owner the right to exclude others from practicing the claimed invention. This discussion focuses on utility inventions, even though much will apply to design and plant patents.

COMPONENTS OF A PATENT

A utility patent contains two main parts: (1) the patent specification, which includes written text, figures, and possibly data that explain the invention, and (2) the claims, which are numbered paragraphs that set out in words the parameters of the invention.

The patent laws require that the patent specification describes the invention in a manner that enables a person of ordinary skill in the art to make and use the invention. Typically, the patent specification contains a background discussion of the subject matter followed by a description explaining the details and advantages of the invention. If necessary, the inventor includes drawings to illustrate further the invention. The patent specification may also include examples of making and using the invention or its components, but examples are not required. In fact, it is not necessary to construct a prototype of an invention as a prerequisite to filing a patent application.

Patent claims must define the invention and are crucial to determining what has been patented and thus, what the patent owner may exclude others from practicing. A patent claim has several components, including an introductory preamble that often identifies the general subject matter of the invention, followed by descriptions of one or more elements or features of the invention. Also known as claim limitations, patent claims establish the boundaries of the invention, which is the subject matter that the patent claim covers.

Claim language is important, and its interpretation is often critical to determining what the patent covers and thus excludes. Claim interpretation can be a complex inquiry, and over the years, federal courts have constructed several rules. Some simple principles include the following: that a claim term is generally given its plain and ordinary meaning as understood by one skilled in the art, an inventor can provide his/her own definition for a claim term, a claim term is understood as it is used in the patent specification, and a claim term should have the same meaning throughout the patent.

REQUIREMENTS OF PATENTABILITY

An invention that leads to a utility patent is often made in the context of solving a problem. Under the patent laws, one can obtain a utility patent for any new and useful process, machine, manufacture, or composition of matter, or any improvement of these items (35 U.S.C. § 101). Examples include a process for making or using something, a device such as an instrument or article that is manufactured, and a chemical composition, molecular structure, or particular formulation.

The utility requirement of section 101 is often considered a low threshold. This section of the patent statue requires that the claimed invention be capable of a credible use, normally a use in industry. A claimed invention must also be novel (35 U.S.C. § 102). Essentially, the claimed invention must be different from the prior art, which substantially is the body of knowledge encompassing what others have done before the inventor filed the application for patent. Finally, a claimed invention may not be obvious to one of ordinary skill in the art at the time the invention was made (35 U.S.C. § 103). The obviousness requirement involves a complex, expansive, and flexible analysis. The obviousness question of validity is answered by determining the difference between the prior art and the claimed invention, and assessing whether that difference would have been obvious from the perspective of a person of ordinary skill in the art at the time the invention was made. In addition, the obviousness analysis looks at secondary considerations. These secondary considerations include the failure of others to solve the problem addressed by the invention, the commercial success of the invention, the existence of a long-felt need for the invention, the licensing and acquiescence of others to the patent at issue, and copying of the invention.

Prior art can come from many sources and includes essentially the body of knowledge and activities known or disclosed by others before the inventor has filed the patent application, regardless of whether the inventor was aware of such information. Assessing the qualification, scope, and content of prior art can be complicated, and as a result, a patent practitioner should be consulted for such an inquiry. While there are exceptions, prior art appears typically in written documents, such as articles and other patent publications, published anywhere in the world. Prior art also includes devices, things, or processes known or practiced in the public before the priority date of the invention (see 35 U.S.C. § 102).

Patent practitioners often counsel inventors to file for a patent application as soon as possible in order to limit the prior art that pertains to the invention. In recent years, this practice has become more important, because in 2011, Congress amended the patent laws to give priority of invention to the first applicant to file for a patent application anywhere in the world. Importantly, an inventor can create prior art against the very invention the inventor is seeking to patent by disclosing the invention to the public, potentially barring the inventor's rights to obtain a patent. These laws are complex, and to mitigate their pitfalls that could cause a forfeit of patent rights, a patent practitioner should be consulted early on in the invention process.

The duty of candor is essential to the proper functioning of the US patent system. Importantly, a breach of the duty of disclosure may render a patent and other related patents unenforceable. The US patent laws impose on the inventor an obligation to disclose what that person knows to be material to an examination of the patent application and the claims sought to be patented. Though much litigation of this concept has occurred, materiality is generally considered anything that a reasonable patent examiner would consider to be relevant to patentability. As with prior art, this area is complex, and advice over what information should be disclosed to the USPTO should be sought from an experienced patent practitioner.

While a search of the prior art is not required to file a patent application, it may provide insight on the patentability and scope of an invention that an inventor can seek. Importantly, the results of any prior art search should be considered for disclosure to the USPTO in accordance with the duty of candor for any related patent application that is filed.

Once issued, a patent is enforceable in federal court and in other administrative proceedings such as the International Trade Commission. In an enforcement proceeding, the patent holder bears the burden of establishing infringement of a patent claim. The patentee shows infringement by proving that each and every limitation of the claim is present in the accused device or activity. The accused infringer has multiple defenses available (Filarski, Patent Defenses, BNA Patent Litigation Strategies). An accused infringer typically asserts that its product or activities do not fall within the scope of the asserted patent claim and, in the alternative, the asserted claim is invalid, because each and every limitation of the claim is either not novel or is obvious in view of the prior art.

PATENTS AND BUSINESS GOALS

While seeking a patent, also consider a business plan. Do this early so that expectations of cost and success are realistic. A potentially patentable discovery may be conceived in a moment of inspiration, or it may result from years of research. In fact, it does not matter how an invention is made according to the US patent laws (35 U.S.C. § 103).

Recovering revenue from a patented invention is a complex undertaking. A successful entrepreneur has certain goals, which are assumed here to include contributing to a field and profiting from that endeavor.

For example, building a company based on an invention usually requires transferring the idea into a product or service that can be sold and used in the marketplace. There are many business models to bring an idea forward. A business plan would include funding and managing for the development, manufacture, marketing, and delivering of the eventual product or service to the customer.

An inventor may not want to invest the time and money to build an enterprise from the ground up. Rather, an inventor may simply want to focus on inventing and allow others to develop a product or service. Alternatively, an inventor may desire a joint venture in which ideas and products are developed jointly with others.

A patent professional should be able to work in accordance with the business objectives of the inventor. That professional may be versed in the ways of getting an idea into the channels of trade. If not, that person likely can help locate those who can advise in this area. Organizations are built on the profession of bringing innovation from the idea stage (concept) to the marketplace (commercialization) (see, e.g., http://www.licensingcertification.org).

In the medical device industry, for example, a device is tested typically in the laboratory and in the clinic before being sold. It also requires governmental review and approval. Patent professionals, business consultants, and organizations can help locate those people who understand this business. There are career professionals who understand the complexities and hurdles encountered in traveling this road to commercial success. Consider consulting these professionals early in the development of an invention.

DILIGENCE AND THE PATENT RIGHTS OF OTHERS

Because a patent gives the holder the right to exclude, but not use, the patented invention, a prudent business strategy for each new product development and commercial introduction is to conduct a freedom to operate study. Also known as a right to use opinion, this inquiry is often conducted or overseen by the patent practitioner. It involves searching for and reviewing patents owned by others to ensure that the new product, if used, would not infringe any other patent. Such diligence is a prudent business practice to follow. The inquiry including any related legal opinions will prove valuable in the event that the product is accused of infringing another patent. These diligence studies, for example, are often cited in patent litigation as a defense against a charge of willful patent infringement.

RELATED INTELLECTUAL PROPERTY LAWS

Intellectual property laws other than patent law should be considered for protecting an idea. These include (1) trade secret—which protects confidential information, trademark, or trade dress—which protects a phrase, symbol, or appearance associated with a particular product or service, and (2) copyright, which protects an original work of art set into a tangible medium of expression. These laws protect an idea in different ways and can lay the foundation for a successful business venture.

Trade secrets, for example, may complement or in some instances provide an alternative to patent protection. A trade secret is anything confidential that gives a business a realistic competitive advantage in the marketplace. Examples of a trade secret include software, a list of ingredients, a manufacturing technique, a method of doing business, or a list of customers or vendors.

In contrast to patent laws, trade secret laws protect an idea without time restriction as long the idea is kept confidential. Trade secrets are not disclosed to the public and do not undergo an examination by any government body.

THE FUTURE OF PATENTS

The current patent laws include several procedures for reviewing the examination process. Congress created two new procedures in 2011 known as inter partes review (IPR) and postgrant review. Both were reportedly designed to strengthen the patent system and, it was hoped by many, to increase the value of the patents granted by the USPTO. Whether these goals have been accomplished is an unanswered question. The IPR proceeding has become widely popular as a defense tool for parties accused of infringing a patent. These parties use the IPR process with some filing multiple petitions to seek a finding from the USPTO that the patent should not have been issued in the first place. The IPR process has had a substantial impact on existing patents, resulting in a substantial percentage of the patents challenged losing at least one claim that was allowed previously by the USPTO. Many IPR cases are undergoing appellate review, and inventors and patent practitioners should pay close attention to their effect on the viability of the US patent system.

IMPORTANT POINTS

- A US patent is akin to a covenant with the United States Patent and Trademark Office (USPTO); the inventor discloses an invention and the government grants the right to exclude others from making, selling, or using the patented invention for a period of time, a type of exclusivity.
- Importantly, a US patent does not give its owner the right to practice the invention that is claimed in the patent.
- The inventor needs to describe the invention in a certain manner to show that the claimed invention IS DIFFERENT from what others have done; it is *very important* to understand all the intricacies of a patent.
- A patent can be bought, sold, licensed, pledged, or used otherwise to obtain funding for the development, manufacture, marketing, and other needs of an enterprise.
- The first step is to retain a patent attorney or patent agent to prepare a written description of the invention and communicate with the USPTO; US patent laws and regulations are complicated, and professional assistance for this task is advised.
- There are three main types of US patents: utility, design, and plant:
  - **(a)** *Utility patents* cover the functional features of an invention that are novel and nonobvious.
  - **(b)** *Design patent* covers any new, original, and ornamental design for an article of manufacture.
  - **(c)** *Plant patents* cover any distinct or new variety of a plant.
- A patent contains two main parts: the *patent specification*, text, figures, and data that explain the invention, and the *patent claims* that describe the parameters of the invention and what the patent owner may exclude others from practicing.
- When seeking a patent, also establish a business plan that includes funding and managing for the development, manufacture, marketing, and delivering of the eventual product or service to the customer.
- A prudent strategy for a new product development and commercial introduction is to conduct a freedom to operate study, also called a right to use opinion; this involves searching for and reviewing patents owned by others to ensure that the new product, if used, would not infringe any other patent.
- Finally, intellectual property laws other than patent law, such as trade secret protection, should be considered for protecting an idea.

Chapter 6

Licensing Medical Devices to Manufacturers/Partnering With Large Companies

Mark Boden and Johnathan Goldstein
Boston Scientific Corporation, Quincy, MA, United States

INTRODUCTION

Medical device companies recognize that *physician entrepreneurs* (PE) are a key component to developing new innovations for health care. PEs are often best equipped to see problems in the current care of patients and to identify and evaluate different options for how those problems might be solved. The PE is a rare breed: a physician who routinely treats patients but still finds mindshare/interests to think beyond existing tools toward a better solution.

It is important for the PE to understand the multiple roles that a large company (LC) can play in turning an idea into a product. With this in mind, a key factor that all entrepreneurs, but especially PEs, need to address is their planned interactions with the large players in the market. In this article, the focus will be on why, how, and when to engage a LC to help develop an idea or to take the finished product to market. This goal will need to be accomplished by a discussion of the structure of LCs, the culture at that LC, and how these considerations will impact interactions with PEs, contact points within the LC, and communication schema to build and sustain a successful relationship. Most importantly, all parties should recognize that the end goal is to share/partner the innovation with the largest end-user population, benefiting patients by bringing better quality of care at a lesser cost to the health care system.

Physicians are also instrumental in understanding the science or underlying physiology behind a disease and treatment. Although basic science can be a starting point for innovation, funding and collaboration for this type of research is outside the scope of this discussion but is discussed in other chapters (state chapters here) in this book.

Medical Innovation. https://doi.org/10.1016/B978-0-12-814926-3.00006-1

CORPORATE CULTURE

Understanding Corporations

It is important to understand the structure and culture of LCs when considering how and when to engage. Working with a LC can be incredibly beneficial, but such an interaction may also be somewhat frustrating for an early-stage company. LCs have multiple internal entities, and the PE may interact with many of them. While sales and marketing functions are the likely face of the company to the PE, there are multiple other functions within a strategic organization likely to provide input at some point during the evaluation process or throughout the collaboration. Business development is responsible for external licensing and acquisitions. Please note: some organizations define business development as the group that identifies new markets and customers, rather than the definition used here. This functional group will have broad expertise in all facets of evaluating a new product, but they will rely on experts from other areas of the company for in-depth assessment during the diligence or evaluation phase. These will include specialists in research and development (R&D), quality, regulatory, clinical, manufacturing, finance, and reimbursement. Most LCs also employ a venture group that provides early-stage investments in areas of strategic interest. R&D teams are responsible for internal product development and work with upstream marketing to set strategic direction. Many organizations have technology scouts, affiliated with business development, the venture group, and R&D, responsible for identifying external technologies and products and integrating them into the pipeline of the LC. Either the technology scouts or R&D will often be the initial touch point to for the initial assessment of interest and will either champion an idea or pass it along to business development or the venture group.

Cultural Differences

LCs can provide all the expertise you need to get from initial idea to a marketed product, but the timelines and interests of the LC are not always aligned with that of a startup. For the purposes of this chapter, startup refers to any technology and affiliated enterprise from patent to revenue-generating corporation. Patent licenses, including for products not yet prototyped, are a common way for new technologies to reach the market. There is no set routine describing how an interaction will go. A few aspects of the cultural differences that can lead to misalignment are described as follows:

1. Organization speed
   a. Initial contact: It may take a LC several weeks or even months to evaluate and respond to a pitch (concept summary and proposal) or questions. Timing will depend on your point of contact, strategic fit, or the need within a company, and the complexity of the proposal.

 b. Full evaluation: A LC has a rigorous evaluation procedure which takes longer and involves more stakeholders than a smaller organization. Multiple requests for information can be frustrating and time consuming for the startup.
 c. Collaboration: When an idea meets a need in the company's existing strategic plan, the PE may be overwhelmed with multiple inputs from all of the disciplines required to move a project ahead. Its worth noting that while this level of attention may not be what the PE would desire, if managed well, this in-depth interaction can be far more beneficial than a simple "check in the mail" with no interaction/critique from the LC.

II. Corporate planning: each LC has a pipeline of products to address a given clinical need
 a. There will be plans for incremental improvements to existing products (Core Product Line).
 b. There will be several "shots on goal" for solutions to expand the portfolio in each strategic area. This may include acquisition, licensing, partnering, and internal development ("Adjacent Product Lines" and "White Space" areas).
 c. Strategic plans and budgets are generally set for a LC on a yearly basis, so it could be months before a proposal is accepted into the coming year's strategic plan, then more time until resources can roll off existing projects to be devoted to supporting a collaborative effort with a startup.

III. More formalized approach to development
 a. LCs have a well-controlled process for product development, while a startup most often has to focus on getting answers to key questions. The speed of startups to develop Proof of Concept and beyond may be commendable, however, fewer resources and less money may leave gaps in the process from the standpoint of the LC.
 b. A startup generally is comfortable with lower levels of verification than an LC, leading to a mismatch in expectations of what will be delivered at each stage in the development process.

In order to manage expectations, it will help for the PE to understand the functions, roles, and cadence of product development within the LC. It is important to stress again that an open eye and mind to the awareness of the culture, structure, product pipeline, and timelines for product development of the LC will smooth the way to a productive relationship and minimize frustration for both parties.

CORPORATE NEEDS—HOW TO GET THE GIANTS TO WANT TO DANCE

LCs with extensive product portfolios consider gaps in the product line and generally look for technologies that "disrupt" current products or meet a need in an adjacent space to broaden offerings. LCs are also generally interested in certain strategic areas, with a focus on disease state or physician call point. When planning a collaboration with a particular LC, consider potential strategic options, whether there is a known gap in the portfolio, or if the product being developed will meet a need in an adjacent market. These considerations and possibilities will help the PE to focus his/her efforts, targeting the best potential partners, and tailoring the message that will be presented to the LC.

There are a number of criteria used when evaluating a concept for investment or in-licensing, including development stage, novelty, strategic fit, protection of intellectual property, solid financial footing, clear ownership, and perhaps most important, the strength and commitment of the startup's team. There is generally a tradeoff of risk versus potential benefit. Often, LCs are looking for opportunities that have been "de-risked," with demonstration of technical feasibility, safety, efficacy, a known reimbursement path, and known clinical and regulatory paths. Strategic fit is a must, while risk level will be assessed against the opportunity size and how well the final product will augment the company's product line in making a decision.

Consider a startup's offering according to Fig. 6.1. It is possible to bring a new product to market in any of these categories. The development stage where an LC is willing to make a substantial investment will be dependent on where the product falls in terms of risk level and strategic fit:

- Most LCs are good at understanding and innovating in the core. They have extensive market assessment and R&D effort dedicated to incremental improvement of a successful product offering. For a startup, considering

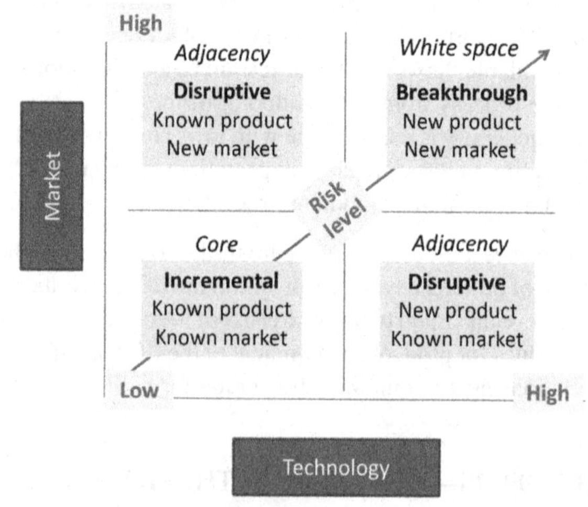

FIG. 6.1 A matrix that delineates between an LC's different product ranges in development (portfolio). Core offerings (lower left quadrant) are at the lowest risk level within currently served markets and known technologies. At a higher risk level, adjacencies are products in new markets or with new technologies. In the upper right quadrant are those products addressing new markets and using new technologies. These would be considered white space or breakthrough categories.

engaging in this quadrant may be difficult, because in many cases, the pipeline is mapped out for several generations. In addition, because the LC has already invested a great deal of time and effort into this product, a new idea will have to expand their market reach substantially in order to justify any external investment.

- In the White Space or upper right quadrant, there is an opportunity for a truly breakthrough product. Because both the product and the technology are unknown, a LC will look for the startup to de-risk the concept by extensive bench, preclinical, and often clinical testing. Often this quadrant would be filled by a large acquisition rather than an early-stage bet.
- Concepts in adjacencies, with either the market or the technology well developed, are often appealing to LCs. There is a moderate level of risk, but good strategic fit, and opening new markets or improving patient outcomes with a new technology is attractive.

PREPARING TO ENGAGE WITH THE CORPORATION

When to Engage

The question of when to first contact an LC is often debated; some thoughts are as follows:

(1) The PE may prefer to prepare a detailed information packet (as described further in the section on Maintaining Momentum in the Relationship: Information Sharing and Communication), including a robust strategy, a clear direction, sharp perspectives of product design, and preferably clinical data to validate the startup's product/idea. The benefit of this approach (of delaying communication with the LC until the product is fully defined) is that the product can be developed independently, and the PE does not need to be concerned about interactions with external parties until ready for a full "show and tell." The LC is also less likely to make a substantial (financial) contribution to a startup at this stage, leading the PE to conclude that there is no return on the effort of engaging early. As described in subsequent paragraphs, this approach is not recommended but may be preferred in some cases.

(2) Earlier engagement can provide the PE valuable input to fine-tune the product/approach, making the product ultimately developed more likely to fit the portfolio of the LC. This approach is valuable, whether the objective is to license the technology at an early stage or to create a startup with the goal of taking the product through to market. The message and what you request of the LC will depend on the preferred exit.

(3) Large corporations take time to ingest the data shared and to come to decisions. In many cases, relationships built over years between PEs and LCs are the ones that create a level of trust, which is the bedrock of the negotiations and leads to optimal outcomes for both sides.

With early contact, the PE will get on the LC's radar, so they can track progress and make decisions about when and how to get involved. The PE may also get feedback on the market, approach, and potential pitfalls from an organization that has invested a great deal of time and effort into researching and understanding the space.

Access Nodes in Corporations

The structure of each company is different, and even divisions within companies vary in their approach to engaging potential partners. There are, however, some common elements or functional roles in each LC (though they may have different titles). LCs often employ specific functions to engage earlier versus later stage startups. We would recommend that the PE identifies and connects with the LC early and finds the right person(s) to champion the idea within the LC. The champion will advocate for the PE's product and make connections to the most suitable personnel at the appropriate time. In the table below are examples of when contact may be made with an LC, the role of the contact person, and what might be asked of the LC at that stage (Table 6.1).

TABLE 6.1 Access Points Versus Stage of Development

Stage	Contact	Potential Ask From LC
I. Ideation/patent application filed	Marketing, R&D, or technology scout depending on company organization	– Verification of interest in the concept, including a letter of support for a grant – Assessment of technical risk – To provide guidance on what information is needed for next steps – Is there a co-development opportunity?
II. Prototype (in preclinical stages)	R&D or technology scout	– General strategic interest – General design input, features – Preferred test methods for verification/efficacy – Funding for further development/preclinical studies – Co-development possibility

TABLE 6.1 Access Points Versus Stage of Development—cont'd

Stage	Contact	Potential Ask From LC
III. Preclinical complete (sufficient development to have a final or near final clinical design complete)	Technology scout, business development, venture group	– Funding for clinical trials – Investment
IV. First in human (or other) clinical trial complete V. Product launch	Business development or venture group	– Investment – Acquisition

Notes: (1) If the preferred approach is to license the technology at any stage, the contact will be the appropriate business development person or the technology scouting group. (2) At any stage, you might approach a technology scout with the intent of informing and beginning to build a relationship, even if there is no specific "ask" at that time.

Other Possible Access Points

As shown in the table, access points to a LC will vary depending on development stage and business model. A few others might be listed as follows:

- Many companies have external-facing innovation portals, where inventors can submit ideas. Ideas coming in through the portal are generally very early, even prior to initial prototype. Submission at this stage will inform if the LC is interested in the concept.
- Direct engagement through known contacts at the company is another route to get feedback. Many PEs rely on sales or marketing to make initial introductions. These groups should be able to make the connection to the right person or persons to assess an idea and provide feedback.
- Only in very rare circumstances can a key opinion leader (KOL), generally identified as a nationally recognized clinical leader, be able to convince the senior management of the LC that the PE has a true novelty without discussing anything regarding the product. This should be assumed to be the exception rather than the rule.

Communication with a LC is best when the PE plans the interaction. Decide on the developmental stage for first contact and find the right person or function with which to open the dialogue at that stage. The next steps will be to put together the information packet that will gain and maintain the interest of the LC and will outline what the PE needs to get from the collaboration.

PREPARING TO SHARE INFORMATION, GENERATE, AND MAINTAIN INTEREST

What to Share?

Part of the process as a PE—at the appropriate stage of development—will be to develop a plan to share information with investors, potential development partners, and representatives of the LC. The PE is and should be understandably concerned about protecting an idea, and it is critical to avoid disclosing details of the invention prematurely. But, in the interest of building a relationship, it is important for the startup to share enough nonconfidential information for an initial evaluation. Just as there are stages of product development, there should be stages of information sharing as the relationship matures. At a nonconfidential level, it is possible to articulate clearly the reason a better solution is needed and which performance gaps will be overcome with the proposed solution. It will be possible to cite references verifying the size of the need. There should be four objectives for this early information sharing:

(1) To discover whether the LC has an interest; if not at this time, what would it take to get them interested?
(2) To build credibility by demonstrating a thorough understanding of the problem and solution and by demonstrating your commitment and passion to developing the solution.
(3) To identify the person with the right attributes to champion the idea.
(4) To whet the appetite of the LC to do the follow-up work necessary to learn more. (e.g., providing this information will enable the company to decide if they want to move forward with a confidentiality agreement.)

The pitch will change and become more detailed and polished as more information and experimental data are collected. Common elements that an LC will look for at various stages include

- A. The BASICS
  - The problem—describe the ecosystem, the problem with the current system/approach.
  - The idea—be able to state clearly and concisely the solution you have identified, key differentiating features.
  - The stage—current state of development of prototype (?).
  - The "Ask"—why are you contacting the LC now? What might they provide?
- B. The DETAILS
  - The team—the strength and experience of your management team and advisory board are perhaps the most important elements in a successful startup.
  - The product/technology
    - The solution is better than what is out there; this is why.

  - Intellectual property—patent application(s) has been filed, and PE has engaged a patent attorney who knows the field and provided broadest possible coverage.
- The Market
  - A brief analysis of the business/market environment to demonstrate your realistic understanding of project potential.
  - Collaborations—You have identified (and partnered) with resources in your ecosystem that will help fill gaps you have identified. These interactions will accelerate the development timeline and will build credibility.
  - Reimbursement—You demonstrate awareness of the payer hurdles in this field, and the understanding why from a health economics standpoint the project is likely to succeed.
- Finance—A robust financial plan which will generate trust when shared.

- C. The STRATEGY
  - Deeper (pre)clinical, regulatory, and reimbursement strategies—Display a clear understanding of what will be required for approval. Even in the early stages, be able to discuss the approval pathway, have insights into the amount of clinical evidence needed for approval, and how the trial will be designed.
  - The work plan—Have a realistic project plan that has been vetted by experts, covering resources and budget. Above all, identify the best target for the solution.

Strategic development plan—Articulate the path from today's status to the desired exit, whether that involves licensing, co-development, or independently taking the concept through to market.

Maintaining Momentum in the Relationship: Information Sharing and Communication

Below is suggested a data-sharing approach that both maintains a healthy, growing dialogue with the LC but at the same time preserves certain details for later sharing, perhaps under confidential disclosure agreement (CDA).

Table 6.2 aims to share with the PE the standard (generic) approach to learning about a company and making the initial connection.

I. BASIC: Initially, a "reaching in" allows both sides to be aware of each other without sharing too much.

II. DETAIL: This is followed by a set of meetings with more subject-matter experts (regarding market, technology, data to date, expectations, etc.).

III. STRATEGY: Once enough data has been shared to justify greater interest, and a CDA has been executed, a deeper understanding of many factors will then need to occur.

TABLE 6.2 Initial Contact

Stage Within Corporation	Information Shared	Key Elements	CDA Required?
I. BASIC: Initial "reaching in" to access nodes	"Teaser"—one pager	Acknowledgement of quality team, IP, market need, direction of solution, stage	No
II. DETAIL: Discussion with experts, post access node	Initial presentation	Details regarding quality team, IP, "pain in market" existence; expansion on solution (the "What") with limitations on the "How"). Startup growth plan and timeline. Ask from corporation	No
III. STRATEGY: Deeper efforts with experts	Detailed presentation	Focus on product features (post-CDA), validation, growth of timeline strategic plans. Ask from corporations (details)	Yes

This step-wise approach is intentionally generalized, but it allows each startup/PE to consider the cadence of the exchange, so that there is an increasing flow of information, communication, and collaboration as the startup becomes more engaged with the LC. Being able to share enough nonconfidential information to engage the LC, without disclosing the invention, will protect both the PE and the LC.

Once the relationship is established, the two elements that the startup needs to monitor/manage now that the project has engendered some interest inside the LC would be:

(1) Information sharing
- Ensure that the LC is receiving data that they request in an organized and succinct manner.
- Share only what is needed, but minimize the amount of items that you are unwilling to share. There might be some elements that you need to object to, but try to maintain these to a minimum.

(2) Communication
- It is perfectly reasonable after advancing to a deeper level of data sharing with the LC to expect that the level of interaction between startup and LC may increase too. The LC—through the access node—understands the importance of this potential relationship with the startup and is cognizant of the fact that the PE has a lot resting on this.

- As such, it will be important for the PE to balance his/her interest in a status update, with the slower (and often more relaxed) approach of the LC to this one startup out of the many in which they are interested. A communication every few weeks—preferably by email—is a good way to ensure that the LC remembers that "I am here" and at the same time is able to continue with their other activities without daily reminders that do not always enhance the relationship.

Maintaining Interest

Achieving key milestones will provide the opportunity for more in depth project reconnects with the LC champion. The LC may identify milestones of interest to them, but you might want to propose your own, including, among others:

- Organizational changes
- Studies of preclinical safety and efficacy
- Fundraising rounds
- Discussions with regulatory and reimbursement agencies
- IRB approval
- First in human trials
- Advanced clinical trials

At each of these milestones, reach out to the LC in order to share the news and to allow the LC to understand how this affects next steps, timelines, and budget. This communication will lead invariably to an improved relationship, and better, faster results when looking for decisions or input from the LC.

GROUND RULES FOR COMMUNICATION

The types of projects that arrive on the desk of the LC fall into three categories:

- Engineer driven
  - A promising technologic solution, often seeking a clinical need to advance this technology toward a product.
- Clinician driven
  - A clinician's perspective regarding an improved solution, without necessarily having a full solution in mind.
- Engineer and clinician driven
  - A perspective that matches a clinician's description of a verified need, with a technology solution for this need.

The optimal project that a PE would bring to an LC would be one in which the technology and business case have been validated by a reputable, bioengineering expert and a business person familiar with the clinical area, respectively.

Relationships developed between the PE and LC communities have led to a communication approach that has at its helm a joint charter of expectations that

TABLE 6.3 Setting Expectations for Communication

Expectations That the LC Will Have	Expectations That the PE Should Have
✓ IP is existent, and reasonably solid; initial pre-CDA information can be shared with no fear of making the startup or the LC vulnerable ✓ Solution adds value to the existing processes ✓ PE is very familiar with the field in which (s)he is inventing ✓ Homework done: know what fields the LC is involved in, and be able to explain how the project would suit them ✓ Honesty: technology is yours, results were as displayed; stage is as stated ✓ The startup might be talking to another LC too. This may be appropriate at early stages of discussions but should not be addressed (flaunted) by the startup	✓ Quick acknowledgement, slower decisions; clarify expectations, and do not expect startup speeds ✓ Professionalism: you will be treated with the respect that every entrepreneur deserves, especially a PE ✓ Confidentiality: the LC understands the importance of not sharing startup data ✓ Conflicts of interest: people inside the LC will exclude themselves if they have an internal involvement (or even IP) that makes them conflicted in a discussion ✓ The LC will have multiple irons in the fire in areas of strategic interest. Due to confidentiality, these will likely not be shared with the PE

one community has for the other. Above we try to capture the (unwritten) laws that govern the expectations of both parties (Table 6.3).

CONCLUSION

Both PEs and LCs are committed to and passionate about providing better solutions to support improved patient care. There are a number of routes to take a concept to market. In most cases, there is a benefit in engaging a LC at a relatively early development stage, just as would be done with regulatory agencies and subject matter experts. When each party understands the other's needs and culture, the steps involved in building and maintaining a relationship, and the ultimate objective, they realize a synergy provided by the speed and flexibility of the startup, and the knowledge and experience of the LC. A collaboration that is managed well will benefit both parties and will go a long way to achieving the ultimate goal of better patient care.

IMPORTANT POINTS

1. Preparation
   a. Understand your product and technology, and how it fits into all aspects of the health care ecosystem.

   b. Understand your potential partner: what their motivations are, and how their culture will impact the relationship.

2. Engagement
   a. Plan when to connect with the partner.
   b. Understand what to share, what to ask from the partner, and how both will change at different stages of the relationship.
3. Relationship management
   a. Identify a champion with the LC, and build a relationship based on mutual respect.
   b. Work with your champion to make sure you understand how the relationship will be managed.
   c. Build momentum to maintain and strengthen the relationship.

Last word: Note that very little is said about the actual technology that is being developed. Although the concept must be based on sound medical and engineering principles, success depends on strong relationships and a robust process when engaging with a number of partners, one of which may be an LC.

Chapter 7

Understanding Venture Capital in the Health Industry

Barry S. Myers* and Charles Robert Hallford†,‡
*Department of Biomedical Engineering, Director of Innovation, CTSI, Coulter Program Director,
†New Ventures, Duke University, Office of Licensing and Ventures, Durham, NC, United States,
‡Adjunct Associate Professor of Finance, Fuqua School of Business, Durham, NC, United States

For most inventors and researchers, taking an idea from inception to market or to a stage in which another company will acquire and develop it will require an outside source of capital. This means (among the many other areas of expertise needed to succeed) that inventors need to understand how capital works. In a broad sense, sources of capital to fuel innovation and translation into the market are classified into either "nondilutive" grants and loans that do not change ownership and "dilutive" equity investments that exchange a percentage of the company ownership for capital. This classification tends to accentuate the idea of ownership percentage as a key goal in corporate development which, as we will argue below, can be counterproductive, impede progress, and increase the chances of failure.

For early development (e.g., proof-of-concept, prototype development, translational research), nondilutive funds may be sufficient to advance your invention and create value. By contrast, later stage, meaning preclinical development [e.g., testing in compliance with the International Standards Organization (ISO), FDA-defined good manufacturing practice/good laboratory practice (GMP/GLP) toxicology experiments], clinical trials, and market launch require substantially more capital than these sources can or are willing typically to provide. The financial strategy you choose to fund your invention may well depend on what is available to you, but more likely is a strategic decision that should be driven primarily by the goal of creating value. Value can be a personal goal of progressing to the clinic or to a license, or it can be concrete and measurable (value = shares of stock owned × share price at the last equity financing). Regardless of the source of funds, choosing a strategy that creates the most value for you, your collaborators, company managers, and investors, should always be your primary goal. When thought of in this way, different sources of capital have advantages (equity accelerates the pace of development, grants

Medical Innovation. https://doi.org/10.1016/B978-0-12-814926-3.00007-3

are nondilutive, etc.) and disadvantages (equity is expensive, grants can be slow and focus on research over value creation, etc.), and the choice requires a clear understanding of how each of these works. In this chapter, we will introduce you to venture capital (VC) with the intension of helping you understand how VCs are motivated so you can increase the chances of acquiring these funds as part of your strategy to create value, to consider the trade-offs carefully, and to bring your ideas to clinical practice.

VC is one source of capital to fuel translation and development [1]. But what is VC? VC falls under the broad category of private equity—investment in companies that are not traded publicly on a stock exchange. VC firms are run by general partners (GPs, or "Partners"), who invest funds raised from investors known as limited partners (LPs). Strictly defined, VC is a financial asset class like public equities, real estate, and hedge funds. In contrast with public equities where shares of an investment are liquid (can be sold for cash at any time), investment in a VC fund is illiquid (cannot be redeemed for cash by the LP until the company is sold or goes public). To compensate the investors for this long-term commitment, LPs expect to earn 2%–5% more than they would in public markets[1] (e.g., against an S&P 500 index fund). If a VC fails to achieve this, they will likely not be able to raise additional funds. Thus, while a VC might have many altruistic goals in social health or economic development, they are first and foremost trying to make money and compete with other financial assets. A typical rule of thumb is that VCs need to see a possibility of at least a 3 × return on any individual investment with up-side scenario, that is, significantly larger.[2] As VC investments in a fund have the potential to be total losses and not return the capital invested, this high rate of return on the successes helps VCs meet the expectations of their LPs. As such, when you talk to a VC, central in your pitch should be to explain how they will make money (3 × plus) working with you. Indeed, a key point here is that your down-side scenario, a scenario in which success is limited, should achieve a 3 × "hurdle rate," and your expected and up-side scenarios should well exceed it.

VC funds range in size from 5 million dollars ($5 MM) to over 1 billion dollars($1 BB) and will have a defined lifetime (typically 10–12 years) during which they make investments, manage those companies, and then must liquidate or "exit" those companies (sell their positions and return cash to investors). As such, VCs with track records will have raised, deployed, and exited several funds. To differentiate themselves from other VCs, an individual firm typically focuses on specific sectors and invests in particular industries, stages of development, and geographies. This approach allows them to build an investment team of experts particular to the area of focus. For example, a firm might hire physicians and scientists for drugs, engineers for devices, and market and finance experts for commercial-stage opportunities. A particular firm might

1. https://blog.wealthfront.com/venture-capital-economics/.
2. http://www.angelblog.net/Venture_Capital_Funds_How_the_Math_Works.html.

be interested in companies that are 18 months pre-IND to Phase IIB ready as an early stage investor or a device investor interested in the revenue stage only. Corporate VCs may have the same or different goals, investing to generate a return or commonly to benefit the strategic goals of the company. A typical fund may make 10–15 investments, so a $100 MM fund might have an investment goal or "bite size" of $3–5 MM initially and reserve $3 MM of "dry powder" for later investment in your company. Thus, as you approach VCs, understand their sector, bite size, age of the fund. etc., and align your approach to meet their needs to increase substantially your chances of raising money for your company. Here, we explore some of the rationales for how VCs do business and how that impacts inventors and entrepreneurs seeking financing.

SECURING OF A VC INVESTMENT: SOURCING AND DILIGENCE

New companies comprise three basic elements: ideas, management, and capital. As researchers and clinicians, we often view the idea as paramount. VCs believe that good management is paramount, arguing that management can save an idea but that the converse is not true. Right or wrong, many VCs also believe that many inventors—as part of the team—play an important advisory role but are ill-suited to serve as CEO due to lack of experience and an unwillingness to leave an academic position to commit full time. Given the challenges that early stage biotech and medtech companies face, VCs often look to experienced managers and inventors they know to mitigate these risks [2]. The good news is that VCs are always looking to build those relationships, building a network of experts that will talk to them about their clinical and research questions. While usually uncompensated, for the aspiring inventor, these conversations will build the relationships you can use later to pitch your ideas and be introduced to potential company managers. Some VCs are willing to start the company and find good managers, whereas others want you to have secured a team.

A VC typically sources (meaning finds) 750–1000 plans/year, investing in two to five of these. To effectively manage that volume, they review and down-select plans quickly, select some for a presentation, and advance a small portion of presentations into a more serious review called due diligence (or "diligence"). The easiest candidates for a VC to decline are companies that do not fit their investment criteria of sector, size, or geography. The second easiest candidates to decline to fund are companies that do not manage to find a personal introduction to the VC through some shared network. A "warm introduction" demonstrates resourcefulness of the management team and provides a degree of validation from the referring party. Because of the value the VC places on management, the chances of securing an investment are reduced dramatically by blindly emailing your business plan.

Should your initial discussions advance into diligence, expect that the VC will do a detailed analysis of your ideas, vet your management team, validate your market, challenge the underlying research and clinical data, and evaluate the competition. VCs will contact thought leaders in the field and poll industry leaders as potential acquirers. VCs vary tremendously in their technical understanding, and therefore, you should look to aim your pitch to their level (always check the backgrounds of those to whom you are presenting). Tone and approach matter; building trust and being a good listener are more important than proving how much you know. As an accomplished or emerging thought leader yourself, it can be a challenge in many circumstances to let others (i.e., your company managers) lead the discussion, yet that is important. VCs believe that even successful outcomes will have challenges. Accordingly, both trust and a willingness to discuss weaknesses and challenges honestly will go a long way to increasing your value to the VC, because the foundations of trust begin at the pitch. If all the elements of diligence are positive, a firm will issue a term sheet—a nonbinding, nonlegal description of the terms of the investment. Typically, the term sheet will include questions to be addressed through additional diligence, a legal analysis of intellectual property (IP), and a condition that the VC will identify successfully other investors who together will "syndicate" to raise all the money you need. In that regard, you need to be prepared and organized to show them all of this in short order, should they take an interest in your pitch.

Valuation—that is, what your company is worth—is driven by market forces, meaning the relative amount of money looking to be invested in and the number of ideas needing money form a market that sets prices. Private market valuations vary considerably, and the best way to have a higher valuation is to secure more than one term sheet. Having more than one interested VC will increase markedly your value [3]. With this in mind, when you are ready to raise money, block out some time on your calendar and run a systematic process with the intention of having multiple VCs in diligence at the same time as opposed to one or two at a time when your schedule allows. Investors will be more responsive if they believe your great idea may end up in another investor's portfolio. In a business context, they will look to you to be equally responsive to email and other communication and having time to do that will play a role in your success. That said, expect this process to take some time, usually months, as an investment is a commitment to a long-term relationship. Use the diligence process to build rapport with your investor and understand their expectations.

THE TERM SHEET

The length and detail of the term sheet varies though, typically, term sheets begin with financial terms of the transaction, such as valuation, the amount of money being raised, liquidation preference, dividends, options, and warrants. They will often include syndication requirements and what constitutes acceptable members of the investor syndicate. This information is then followed by terms describing the corporate structure, which include composition of the board of directors, the voting rights of the various classes of stock, a variety of terms that govern process and decision-making in the event of an acquisition offer (e.g., drag-along, conversion, participation), and protective provisions (redemption, preemptive rights for a future transaction, antidilution, etc.). The term sheet will then include general terms, such as expiration of the term sheet, and that the term sheet is nonbinding and likely confidential. Finally, the term sheet will include all items of remaining diligence and any other conditions that must be satisfied prior to closing.

Understanding each of the terms in the term sheet is beyond the scope of this chapter. That said, however, it is important to understand what each term in the term sheet is intended to do. Like the practice of medicine, investors have their own jargon, and you need to understand it (or have experts and counsel that do understand it and can explain it to you). There are lots of great resources online to assist you with the term sheet (e.g., Investopedia.com), but evaluating the particular terms of a proposed term sheet benefits substantially from prior experience and in depth awareness of the current market. Experienced legal counsel is invaluable in this process, because typically, these experienced experts have seen multiple deals and can provide advice as to what terms and values are fair in today's market.

As important as understanding the definitions of the terms, however, is to understand the implications of each term. Financial terms are there for a reason—they align interests and are a way for individuals to insure the commitments they made at the time of presentation. Consider a 1× liquidation preference. This term prescribes that in the event of the company being acquired, investors receive an amount equivalent to their investment (often plus dividends and interest) first, and then any remaining proceeds are divided based on a percentage of stock ownership. On first blush, a term like this might be considered unfair. Why do investors effectively get more than their ownership percentage? The answer is two-fold. Consider a hypothetic company that raises $1 MM ($1 MM is short for 1 million dollars) agreeing on a premoney valuation (the value of the company before the transaction) of $2 MM. If, the next day, the company were to sell for $2 MM, the initial shareholders would have effectively converted $2 MM in theoretic paper value into $1.34 MM in cash. While this represents a decrease in value, the stock holders of the company might consider this a substantial win, while the investor would take an immediate financial loss of $0.33 MM. With a 1× liquidation preference in place, however, the investor would receive $1.33 MM and the company only $0.67 MM. This approach protects the investor in two ways: (1) it engineers a positive return on investment in many downside cases, and (2) it decreases incentives of other shareholders to accept acquisition offers for modest returns. More to the point on fairness, the company has asserted that they will do work, advance the innovation to the clinic, and be worth tens of millions of dollars when they proposed the investment. Should they make good on their promise and sell for $100 MM, the $1 MM of liquidation preference becomes less consequential. Thus, this term is the company's way of backing its optimistic claims about future success and insuring it will not settle for a small victory at the expense of the investors. Other terms, like stock options that vest years in the future, ensure ongoing effort for success by the management and the founders. Terms such as these can also be used to deal with differences in valuation. If one group believes the firm is worth more because it will achieve certain milestones, and the other disagrees (and has a lesser valuation), adding options that will be issued only with successful realization of the milestone allows both sides to be satisfied at the time of the transaction for an event that will occur in the future.

Finally, as you look at the term sheet and while you will argue every term individually, try not to negotiate each term without considering the context of the other terms. Ask the larger question, "Is this a good term sheet?" How does it reflect your interests and goals? Doing so will allow you to negotiate the individual terms more effectively.

PLANNING FOR SUCCESS: VALUATION, CONTROL, AND THE INEVITABLE SETBACK

As we alluded to earlier, valuation is determined by market forces. If there are lots of buyers (VCs), then valuations go up, but in contrast, if there are lots of sellers (inventors and companies needing cash), then valuations fall. If your company is in trouble financially for any reason, and buyers learn this, the valuation will decrease. Because real revenue is often years away, mathematic determinations of valuation (like discounted cash flows) are not very helpful in setting valuation, and thus valuation is linked usually to accomplishing regulatory or other commercial milestones. In today's market, a concept based on research might be worth $2 MM, a lead molecule 12–18 months from a FDA Investigational New Drug (IND) is worth $5–10 MM, a completed Phase I drug is worth $15–30 MM, and a Phase IIa compound with some nonstatistical hint of efficacy is worth $70 MM.

Let us examine the consequences of these large step-ups in valuation for clinical stage assets in the following hypothetic scenario: A company raised successfully a Series A round of $4 MM for your very early concept that everyone involved agrees is valued at $2 MM (investors provided $4 MM in capital for 67% of the company). The valuation of the company (postmoney value) is now $6 MM, and with this money you and your company advance successfully to a first-in-human trial. The company and the investor bring in two more Series B VCs (the initial investor is able to contribute $2 MM more to the new round, and the other two larger funds contribute the remainder) and collectively, they raise $12 MM in cash with a new (prior to investment or "premoney") valuation of $12 MM (up from the prior $6 MM postmoney value) bringing the total firm value (after investment or "postmoney") to $24 MM (see Fig. 7.1). The firm is advancing to a Phase IIa study with this money, and your idea has an experienced team of managers to ensure manufacturing, accounting, legal, IP, human resources, tax, insurance, fundraising, quality, regulatory, reimbursement etc.

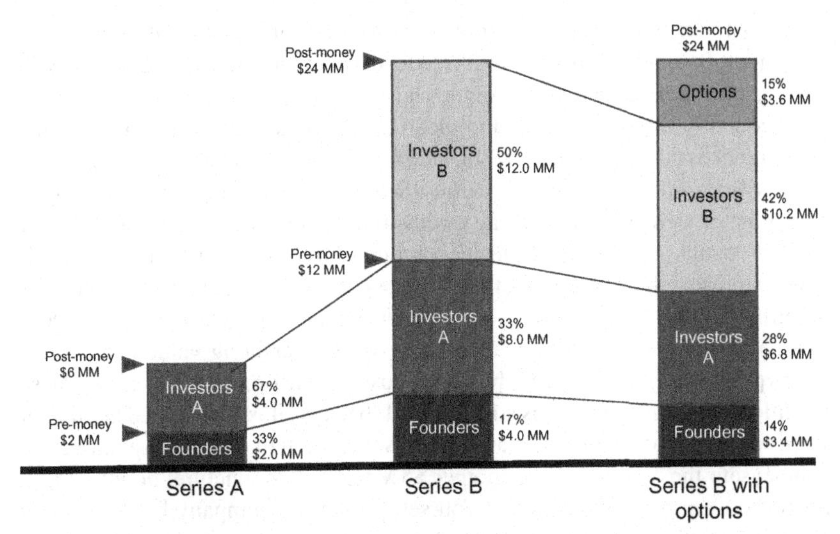

FIG. 7.1 The effect of investment and milestones on ownership and value showing the substantial dilution of your stock in the company despite the increase in your overall value related to the clinical success and VC investment.

are attended to. To provide incentive for management to work diligently toward your company's success and not just draw a salary, you and the VCs allocate 15% of the equity to them as options. In the simplest structure (and one most favorable to you), you now own 14% of the company. The board of directors is dominated by the three investors (many VCs are "active investors" and insist on a seat on the board to steward their investment), the CEO, perhaps you, and one or two independent members. The result is that you have lost control of stock voting and the board voting and suffered substantial dilution!

Realizing that success means losing control and substantial dilution is not an easy thing to come to terms with and illustrates that equity capital is expensive and comes with certain obligatory strings attached. But, consider what you have gained in exchange. You now have a professional team focusing 100% of their time on developing your idea, you have investors who can fund the current round, participate in the next round, and assist in the eventual exit for your company, and your idea is in clinical development. In addition, you are able and encouraged to publish the company's data and present it nationally, the company may subcontract research to your lab, work with you to acquire federal funds, or hire you as a consultant, and what was a $2 MM value in founder stock a couple of years ago is now $3.4 MM, so your valuation has increased considerably. Remembering your valuation priorities, this outcome can be a great success for your career and your company. In that regard, while it is important to always protect your ownership percentage as much as is practical, 100% of $2 MM (your starting point) is substantially less value than 14% of $24 MM.

This illustrates that value creation over ownership interest should be your primary goal. By contrast, if your goal is to lead a company and control it, you will need to consider other funding strategies.

Frequently though, biotech and medtech companies suffer setbacks on the way to success. Indeed, the initial business plan that you wrote usually diverges quickly from actual corporate activities. Setbacks are the norm and can include a myriad of delays—clinical site acquisition, patent litigation, the FDA or another regulatory agency requiring resubmissions, delays by suppliers, challenges in research, slow contracting, changes in management, clinical setbacks requiring retrofit or reformulation, or slower than expected enrollment. Setbacks invariably cost the company money without creating value. Imagine in our hypothetic scenario in which a delay costs $5 MM more to reach the Phase IIa milestone. The valuation is still $24 MM (because it is determined by market forces and not cost to get to Phase IIa) so someone will have to lose money to compensate the VC paying in another $5 MM. At this juncture, how you as a founder/inventor have positioned yourself within the company becomes critical. Are you aligned and deeply valued (i.e., needed) by the company going forward? Or, are you viewed as difficult, argumentative, and focused on issues and development activities that have limited impact on company value? Regrettably, it is an easy situation to be in. Perhaps you have a new idea, and you know it is even better than the clinical stage program built from your first idea. As a scientist and clinician wanting the best for patients, developing a better product makes perfect sense. The company chooses not to bring it to the clinic realizing that abandoning their clinical program for the new preclinical program will decrease the firm's value by more than 10 ×, and the cost of bring both forward exceeds the value created. Under this scenario, your interests and the interests of the company are no long aligned.

Faced with setbacks, the stock price will likely decrease for the next round of financing. Investors who agree to put in more money in the company will recover some if not all of their losses, whereas investors who cannot support the company may suffer very substantial losses (owing to "pay-to-play" penalties). If management is considered capable and not at fault for the setback, they too will be made whole by granting them new options or other financial incentives such as carve-outs to keep them in the company. As a valued part of the team, you will also be made whole (e.g., via issuing options, etc.). By contrast, if you are not considered valuable to the company, your financial loss can be dramatic. How then do you ensure that you are valued by the company?

Working through the remaining diligence, securing a syndicate, and successfully closing a round of financing is a cause for celebration and a major accomplishment that sets the stage for considerable growth and value creation for your company; however, as you plan your current financing, be certain to position the company for a successful follow-on round. Understanding the answers to some key questions can help you succeed. For example, does the current syndicate of investors have sufficient reserves (called dry powder) to contribute to the next round? Do all your investors have the same expectations for a time to exit, or are some in older funds that will want to exit earlier? Will the new valuation and board structure be attractive to new investors in the next round? Does the current round of financing achieve creation of real-value milestones and allow runway (time) to raise additional capital? Indeed, looking all the way to a successful exit via a public offering or acquisition and working your way back to your current position should be a part of each financing. For example, if a Phase IIb asset is expected to have an initial public offering (IPO) valuation of $150 MM based on current trends in the public market, then your company needs to arrive at that stage with a valuation that mezzanine (pre-IPO) investors will find attractive. While often an act of necessity, raising funds without attention to future financing may leave you with a lot of work to do and little chance of success despite realizing your milestones.

THE ROLE OF THE FOUNDER/INVENTOR

Biotech and medtech companies grow in distinct phases: preclinical development and regulatory preparation, clinical development, and sales and revenue. In the same context, inventors, researchers, founders, and managers are often best suited to one of these phases. In simplest terms, the attributes of a CEO who can measure and drive sales growth differ substantially from one leading a research team to establish a potential lead molecule. As a founder/inventor/clinician, understanding where you can provide leadership to the company versus where you can contribute but lack operational experience is important to establishing a long-term relationship with your company as it grows through these phases. There are several key roles that you can uniquely fill. These opportunities include sharing your clinical and research acumen by networking the company to thought leaders in your field to help build its Scientific Advisory Board. Publishing and presenting research related to your invention are especially valuable. As you advance to the clinic, the FDA will take considerable reassurance if peer-reviewed journal articles support the underlying mechanism of action and/or scientific hypothesis. Reimbursement decisions by insurers, while perhaps many years away, are driven by compelling scientific evidence including preclinical studies. For companies in the market, the FDA regulates what they can communicate based on the approved or cleared label. The company can distribute publications freely, and therefore, your published work, presentations, and media can support sales for revenue stage companies. Furthermore, as you continue to invent, your new inventions can serve as new licenses to the company to build a pipeline, expand the clinical indication, or provide extensions of the product line via creation of new IP and important features. Understanding these opportunities and their value to the company relative to their lead program will go a long way toward building a sustained valuable and aligned relationship with the firm.

CONCLUSIONS

As a final thought, realize that capital markets change quickly. Over the last 10–15 years, there have been no fewer than three cycles of boom and bust in the VC business. During the real estate crash, VC capital was hard to find, because investors reserved more cash to protect their existing companies. After the recovery, there was consolidation, creating a few larger funds with larger bite sizes. With only so many late phase clinical trials to fund, we saw the emergence of seed investments worth tens of millions of dollars in platform technologies but with substantial technical risk, many years to IND, and small orphan markets invigorating clinical problems that were considered uninvestable for these same reasons only a few years prior (e.g., viral delivery of a gene to treat orphan disease). Capital from Asia and corporate VC emerged as a very prominent source of equity capital. Charitable disease foundations began making equity investments, and federal grant money began supporting more IND/IDE-enabling studies. Most recently with the now 8-year-old run-up of public markets, wealthy individuals (angels) and groups that previously did only small seed rounds now represent a very real source of investment that can take innovations into clinical development. All of this has made for a very robust environment for inventors/entrepreneurs today; but of course, that could all change tomorrow.

What does this all mean to a clinician researcher/inventor? Devote a small portion of your time to understanding investors and their goals, and become a valued founder and consultant so that when you next invent, you will have VCs and other seasoned investors who can vet your ideas, shape your strategy, and perhaps give your company money with the goal of getting your innovation into the clinic. Realize that invention and development are largely learned experientially, so spending time with mentors and experienced colleagues will be invaluable as will the realization that this may be your first invention that will go out to raise VC funds but not likely your last.

IMPORTANT POINTS

- Sources of capital to fuel innovation and translation into the market are classified into either: "nondilutive" grants and loans that do not change ownership, or "dilutive" equity investments that exchange a percentage of the company ownership in exchange for capital.
- The financial strategy to fund your invention should be one that creates the most value for you, your collaborators, company managers, and investors and should be your primary goal.
- VC can fuel translation and development; VC is a financial asset class like public equities, real estate, and hedge funds. Investment in a VC fund is illiquid (cannot be redeemed for cash by the limited partner). Likewise, a VC investment in your company (i.e., the investor in the VC) is illiquid until the company is sold or goes public.

- As a general rule, VCs need to see a possibility of at least a 3 × return on any individual investment, with an up-side scenario that is substantially larger.
- VC funds range in size from 5 million dollars ($5 MM) to over 1 billion dollars ($1 BB) and will have a defined lifetime (typically 10–12 years) in which to invest those funds, manage and grow the companies invested in, and exit those companies.
- Should your initial discussions advance into diligence, expect the VC to do a detailed analysis of your ideas, the "value" of the company, as well as vet your management team, validate your market, challenge the underlying research and clinical data, and evaluate the competition.
- Term sheets begin with financial terms of the transaction, such as valuation, the amount of money being raised, liquidation preferences, dividends, options, and warrants. They also describe the corporate structure, define the decision-making in the event of an acquisition offer, as well as detail the protective provisions.
- Understanding each of the terms in the term sheet is paramount and experienced legal counsel is invaluable in this process.
- *The role of the founder/inventor:* The skills that make for a successful research clinician or scientist often differ significantly from those that make a successful start-up CEO; understanding where you can provide leadership and input to the company versus where you lack operational experience is crucial to establishing a long-term relationship with your company as it grows.

ACKNOWLEDGMENT

This work was supported by the generous support of the Duke Innovation and Entrepreneurship Initiative (https://entrepreneurship.duke.edu) and the NIH grant 5UL1TR001117 in the Duke Clinical and Translational Science Institute.

REFERENCES

[1] Mazzocchi RA. How does an inventor find an investor or partner? Raising funds to start a company. Surgery 2017;161(5):1183–6.

[2] Makower J. Inspiration, perspiration, and execution: an innovator's perspective. Surgery 2017;161(5):1187–90.

[3] Henry GF. Funding innovation: moving the business forward. Surgery 2016;160(5):1135–8.

Chapter 8

The Process for Innovators/ Founders to Raise Capital to Start a Company☆

Rudy A. Mazzocchi
ELENZA, Inc., Roanoke, VA/Zürich, Switzerland

As discussed in many of the other chapters in this book, an overarching theme of the inventor is to develop medical innovations that should not only incrementally optimize outcomes, improve procedural profits, and/or marginally decrease the risks to the patient, but also address the rather lofty goal of addressing an "unmet clinical need." To accomplish these related goals, substantial financial investments will be necessary. Although investing your personal savings to initiate the development of your novel technology will show commitment and help others to accept your personal optimism of success, the real validation of your invention or innovative idea that drives the process of clinical validation to commercialization will eventually come from the participation of sophisticated, experienced, and financially astute investors with substantial capital resources. Once the inventor believes that his or her discovery represents a true innovation, the new, less experienced inventor needs to then ask several key questions before going through their savings to finance their initial prototype:

1. Does the discovery address a truly important and "unmet" clinical need, or does the innovation only serve as an incremental improvement or enhancement of a current therapy? In other words, is the discovery a product or procedure worthy of building a company?
2. In terms of the extremely important clinical "first-in-man" assessment, what is needed to generate the "proof of concept" and then to establish evidence-based "clinical validation"?

☆ Adopted in part from Mazzocchi RA. How does an inventor find an investor or partner? Raising funds to start a company. Surgery 2017;161:1183–6.

Medical Innovation. https://doi.org/10.1016/B978-0-12-814926-3.00008-5

3. What are the inventor's personal goals: to build equity value, generate licensing income, and/or to really develop, market, and commercialize the innovation for global applications?

Before proceeding to the next step beyond the discovery process, these important questions need to be considered and then answered to the satisfaction of the inventor *before* proceeding on to the development of this innovation. If the discovery truly has merit and can justify the time and energy needed to form a company, a substantially large commitment of financing will be required. To proceed further, two additional questions must be both understood and addressed: (1) what is the landscape of the intellectual property (IP), that is, is the technology protectable? and (2) is there a viable, substantial global market? These initial questions may require the input of a qualified and experienced patent attorney to perform an in-depth patent search to identify if someone has already filed the same elements of the invention (i.e., "prior art"), and/or if the invention/innovation justifies filing a patent with a new set of marketable claims to protect the invention.

In order to attract experienced and qualified investors, an understanding of how the innovation can impact the global market is instrumental in determining its initial value. This market intelligence needs to address the potential revenue model, the pathway for regulatory approval both domestically and internationally, and whether reimbursement/third-party payments can be expected from *each* major country. Your operating plan and budget requires insight into all these factors and will serve to support your assumptions and address your challenges to achieving success. Once you have developed your strategies concerning technology development, mapped out the market potential, and addressed your IP position, the next step is to recruit an experienced management team with a proven track record to execute the business aspects required to market your innovation from concept to commercialization. Accomplishment of preestablished goals will truly depend on the expertise and insight of your management team.

IDENTIFYING THE RIGHT INVESTOR

Obtaining grant support for the initial funding of your discovery from sources like the NIH, the program of Small Business Innovation Research (SBIR), the Defense Advanced Research Agency (DARPA), or other federally funded projects designed to jump start companies, is limited and time consuming. Although these programs can be important in the early phases of research and discovery, these avenues should not be considered as true "investments" as such, because these sources of funding do not provide for a sustained, liquidated return to qualified investors. These and similar sources of start-up capital rarely provide a guaranteed, long-term financial commitment of sustained funding through the entire process of commercialization and/or to achieving profitability (i.e., "positive cash flow"). The more conventional financing includes separate classes of private and institutional funding, often sequential, and each having their own distinct set of terms, conditions, and criteria of "valuating" the innovation. These more business-oriented investors often segregate into several distinct groups: (i) family and friends, (ii) institutional or venture capital, (iii) private equity funds, and (iv) strategic/corporate investors. While exceptions exist, each of these groups has unique and insightful criteria for their financial commitments. These criteria address their predefined interests associated with the current developmental stage of your "idea," the global market, the various terms of their investment, and considerations of the timing of a potential exit or liquidation from their commitment. It is important to understand that the most experienced and successful investors will themselves also research your claims of market potential, address their own professional perspective of the possible IP protection, and assess the ability of your management team to deliver on the defined plan. Each of these investor groups may have an impact in your development, but the individual goals of each need to be fully understood before committing to their requirements and accepting their capital.

Family and friends: This source of start-up funding may prove important in the initial phases of courting a major funding source. Most follow-on, experienced investors want to see some commitment by the inventor of a willingness to invest personal funds and sweat equity in the project which goes beyond the contribution of just the IP or the intellectual capital. While these contributions are crucial commitments, alone they are not enough. Financial input from the inventor's pocket need not be the primary or the major source of initial start-up funding, but some objective sign of commitment (evidence of a personal belief in the project) needs to be shown to convince the investor of personal "skin-in-the-game." This usually means either an initial financial outlay or a very substantial and sustained time commitment. It is important for the inventor to understand this *before* seeking substantive capital investments from others. Along these lines, appropriate avenues of investment can include some degree of capital contributions from family members and friends as buy-in to the idea, primarily because this type of investment capital is at the greatest risk and is viewed by the investor as a personal commitment to friends and family to whom the inventor is accountable. Future major investors realize that there is no guarantee from these initial sources of funding that additional follow-on capital support will be provided.

Venture capital: The world of venture capital has changed tremendously these last two decades as influenced by the changing arena of finance. The once classic strategy of "start-up capital funding" as needed in the early stages of a company has been largely abandoned by many venture capital funds. Many of these investors have actually left the true business of venture funding of innovations in medicine and health care. The remaining venture capital investors now expect higher returns by scrutinizing premoney valuations, often tranch their investments according to specified milestones, and demand voting rights and higher liquidation preferences to protect their investment before committing their financial support. This places more demands on the inventor to perform and provide a deeper assessment of the potential market opportunities, carry out an evaluation of noninfringement and patent protection, develop or at least identify a team of qualified and experienced associates to help manage the initiative, and draft a formal business plan with operating goals, achievable milestones, and insight into how the financing will be used to develop the innovative idea (i.e., "use of proceeds"). This intense and demanding approach by venture capital groups serves to minimize their risk and often stipulates a staged or "tranched" investment based on the achievement of a specific set of objectives. These expectations and demands by early-stage venture capital funds has been a major change in the current scene from the past, where much of this deep assessment and development was done once financing was secured.

Private equity funds: In the past, the concept of private equity funding was usually for market expansion, to provide capital for mergers and acquisitions, or to provide bridge financing to initiate the public offering of a company. But currently with the recent disappearance of the more classic financing opportunities for early stage venture funding, several of the private equity funds have changed their overall strategy to provide development-stage investments as part of a longer term "roll-up strategy" (i.e., to possibly combine two or more related businesses) or to "bolt on" another company or idea to supplement a missing key element in the original company that would enhance the market share and/or strengthen the product portfolio. As might be expected in today's competitive climate, these investors also require more detailed, upfront work by the inventor/innovator in terms of market analysis, patent protection, and defined use of proceeds. These private equity firms, however, will often support the longer-term financial requirements of the company and potentially become more of a "partner" with the inventor and the management team.

Strategic/corporate investors: While the idea of attracting a market leader in the industry who shows strong interest in your idea, your prototype medical device or your innovative "concept" can be really exciting to the inventor, there are several potential disadvantages that must be acknowledged. This stage of developing and commercializing your innovation with the "collaboration" of a corporate investor may prove to be the most challenging process in your recruitment of an investor and the ultimate development of your idea. As the inventor, one needs to be keenly aware of the ultimate goal(s) of the big, established corporate investors whose parochial interests may not be totally aligned with those of the inventor and may be destructive to the management team and initial investors in the longer term. The approach of many corporate investors may dramatically differ from that of the more classic venture capital investors whose plans, processes of involvement, and expected outcomes are very clear and often serve to align more accurately with the inventor to help to promote the development of the technology to reach the goal of optimal commercialization. With corporate investors, the inventor must develop and maintain realistic insights into the ultimate goals of the corporate partner. There is often potential competition with the corporate investor and their existing products or those secretly under development. The ultimate goal of the corporate investor may be to offer financial support to gain access to a license of the patented technology or acquire proprietary knowledge of the new innovation from the inventor. In addition, they may have alternative motives in how the innovation is brought to the market and may seek to control or even prevent the idea from impacting the corporation's long-term business objectives. This possibility by certain industry moguls often is not based on any interest in developing a truly unique advance in health care but rather based on the strategic financial and economic issues of their company to maintain the market share in their field of interest. Because of these potential issues, inventors need to remain open-minded and possibly even consider this ostensibly exciting source of financing as a "last resort."

PREPARING FOR INVESTMENT

Once the decision is made to make the step up to the next level of financing, the inventor will need to commit time and diligence to prepare a detailed "investment package" for the potential investor. This important step requires many of the factors and considerations described above when soliciting an experienced, qualified investor. You may wish to involve close friends and relatives who have provided initial seed money and who understand just how you plan to spend a portion of their retirement funds in exchange for a handsome future return. Your insight into the depth and type of the investor's involvement will determine whether this package will be concise and sufficient enough to secure their investment, or whether it is necessary to provide a more comprehensive, long-term strategic plan based on the complexity of the market development of the invention. The following categories are essential in preparing this investor package:

The executive summary: This summary should be an overview of the technology of the invention, and its improvement, replacement, or introduction of a totally new therapy compared to the current standard of care, along with a clear, concise, and convincing description of the potential impact on the current market. In addition, it is necessary to provide a profile of the patent landscape and the estimated capital needed to reach the necessary milestones of the business to achieve the next series of inflection points in the company's increased valuation. This executive summary should also address the strengths, track records, and expertise of the management team, including their biographies to reveal their focus on the type of innovation involved and prior experiences in this specific field. Important in this summary is a description of the proposed general structure of the legal entity in which the inventor is seeking the capital investment, such as an S-Corporation (i.e., a sole proprietorship), an LLC (Limited Liability Corporation), or a C-Corporation (i.e., a formal incorporation in a designated state of jurisdiction). Experienced legal advice should have been solicited *beforehand* to help to determine which legal entity is best for your ultimate goals and needs.

Projected budget: This part of the proposal usually involves a spreadsheet and summary chart that itemizes the costs involved with driving the technology and final product forward through the initial steps of "proof of concept" to some defined form of "clinical validation." The importance of this section of the proposal is to provide assurance that the proposed funding is adequate to support the safety and effectiveness in human subjects. Although many first-time inventors tend to de-emphasize or underestimate the expenditures in an attempt to magnify potential profits, a qualified investor will be critical in their assessment of these timelines and related budgets. It is therefore in the best interest of the inventor to "reasonably" *under-promise* and *overdeliver.*

Capitalization table: Based on the issuance of founders' shares and previous investments (including family and friends), the capitalization table should be crafted to convey the current disposition of individual ownership. To ensure proper alignment with the founders, the management team, and equity investors, it is essential that this part of the document also include an allocation of proposed stock options grants to incentivize the expansion of both the management team as well as advisors and potential other directors as deemed appropriate. This important clause, called an "ESOP" (employee stock option plan), acts as a carrot for high-caliber new hires and often represents as great as 15%–20% of the fully diluted share basis of the company. Occasionally, this capitalization table will also show the equity position of new investor(s) and the effect on dilution of the existing founder's shares based on a proposed valuation of the company.

The private placement memorandum: This legal document requires the expertise of a corporate attorney who is charged with aiding (and thereby protecting) the less well-informed inventors by helping to identify all the risk factors associated with this investment. This memorandum may be of interest to private investors but is less important and often not needed when soliciting funding from venture capital or private equity investors. This memorandum is often considered to "wrap-around" the executive summary, budget, and capitalization table.

The personal presentation—the "pitch deck": In order to achieve a financial commitment, all qualified investors will require an in-person presentation. This type of "audition" allows the investor to not only observe how well the inventor presents and understands his or her innovative idea but also to determine the level of commitment to the project and overall passion of the inventor to succeed. These "pitches" are extremely important, need to be convincing, and can make or break the investment opportunity. The inventor and/or one of his/her partners should walk the prospective investor through the technology, the market impact, the operating plan (including budget and timelines), and an overview of proposed management. By using visual aids (e.g., a powerpoint-type presentation incorporating videos, cameo appearances by patients and/or clinicians, or other reputable representations), such presentation can be considered a virtual art form of marketing; importantly in this regard, even the colors, flow of the pitch, and other imaginative additions can serve not only as reflections of the sincerity and skill-set of the presenter, but also the imagination and dedication to win over the investor. This presentation should include a very concise set of discussion points on each topic highlighting certain informative and critical aspects of each but without overly repeating the content of the documents sited above.

TARGETING QUALIFIED INVESTORS

Fundraising for a new venture to develop a novel innovation is a very tedious, inefficient use of time for inventors who usually lack the appropriate skills and expertise needed for such an important activity. Due to the importance of this critical aspect of funding and driving the adoption of the innovation, most new and insightful inventors are wise to hire a professional fundraiser, advisor, or even an experienced *pending* Chief Executive Officer with the charge of attracting and obtaining this first substantial capital investment. In certain fields, there is a known network or consortium of funds that are or have been involved actively in capital investment in similar areas of innovation known to individuals who have raised capital previously in a specific industry. Identifying and targeting these potential investors will be aided by an experienced fundraiser and/or advisor.

In the event the inventor(s) desires to raise the capital on his or her own, several proven avenues of research are available to identify the most likely venture capital, private equity investors, or corporate investors.

Prior transactions: One should start with a search for companies in a specific sector of health care that have received substantive funding or that have been recently acquired. Press announcements can be a perfect source to help identify active investors in specific markets or who support particular technologies. Reviewing investor websites may help identify the individual (partner) in the group who was responsible for evaluating and actually managing the investment or purchase of a related or relevant venture. These are the best individuals to target with your initial pitch.

Medtech/venture conferences: Regional and National conferences are often sponsored by investment groups who support a specific sector of the medical/health industry and use these conferences to identify additional opportunities for investment. Conference websites will contain potentially valuable insight into the individual sponsors as well as the moderators of certain sessions who are well-recognized as potential lead investors. Moreover, by attending the presentations, valuable information important to these investors may be evident in some of the presentations at the conference to revise and enhance your pitch deck. Equally important is visibility of yourself and your idea. Any interaction you can have with such investors will only help your recognition and potential presentation of your ideas in the future.

Approaching corporate investors: Several leaders of the health care industry, such as Novartis, Medtronic, Abbott, and Johnson & Johnson, have established corporate venture funds managed separately but which are relevant and synergistic to the corporation's overall operations. Other medical venture funds, such as those managed by the Mayo Clinic Foundation and the Cleveland Clinic, function less as true corporate investors and more as "collaborative incubators," albeit often with less potential funding available to help to achieve commercialization. Other large corporations use their divisions of business development to manage their strategic investments. It is often possible to better understand the objectives of these existing divisions by accessing the company websites or by initiating a conversation through the company's local sales manager or regional distributor.

CLOSING THE DEAL

It cannot be stressed enough that you should NOT try to go it alone. While preparing for and seeking capital investments, unless you are a seasoned and experienced founder with a successful track record, you will need to build strength and diversity in your team by adding knowledge of the business to supplement the credibility and experience that the founder(s) may lack. Although the founder/inventor(s) may understand the invention better than anyone and may think that they sufficiently know the potential market, the experienced qualified investors will expect them to identify an experienced management team to supplement the lack of business knowledge or experience of the initial founders of a company. In the world of medical technology, the practice of medicine and the clinical science provided by founders are a very important and crucial expertise, however, in the world of hospital administration and institutional business, other considerations may drive investor interest. The current financial pressures invoked on these entities now and in the immediate future, the arena of regulatory affairs, and future projected reimbursements from governmental and private insurers, are all equally important to the eventual end-users, for example, hospital and institutional CFOs and the value analysis committees. Members of the management team who can best address these key elements may be the company's legal counsel, a financial manager, or perhaps someone with expertise in market development. This type of a multidisciplinary team gives the investor more confidence before agreeing to fund your project.

Finally, one of the most important and challenging elements of funding your invention is to determine its true "value." To drive a novel, innovative concept through all the phases of development to its commercialization requires a great deal of fortitude. A vision of success and/or fame, the definitions of which vary for each individual on the team, and the insight to acknowledge and accept failure along the away without wavering in such a quest are key to ultimate success. If one places too high of a value and emphasis on early-stage innovation or abandons suggestions and input from experienced advisors, proper funding may never become a reality.

Remain realistic about the multiple risks and challenges ahead, the amounts of additional follow-on capital required, and any ongoing disruptive (competitive) developments by others in the field that may become unforeseen hurdles. Choose your investors wisely. Be reasonable and realistic with your expectations regarding valuation and outcomes, and accept and learn from both your successes and your failures along the way.

IMPORTANT POINTS

- Before starting this lengthy process, ask yourself: Does this discovery fill an unmet need, what clinical validation or "proof of concept" is necessary, and what is your ultimate goal in the process?
- Early sources of start-up capital can help initiate an idea but are not designed to be reliable avenues of sustained funding.
- More viable sources of long-term funding include: (i) institutional or venture capital, (ii) private equity funds, and (iii) a strategic/corporate investor.
- Beware the motives of the big corporate investors; their goal may be to simply control this (competitive) technology and not to aid in its commercialization to avoid conflicts with their ultimate corporate business objectives.
- Once you decide to seek the next level of financing, a detailed "investment package" is needed and must include: (a) a concise and well thought-out executive summary, (b) a projected budget, (c) a capitalization table, and (d) the private placement memorandum.
- Be prepared to devote considerable time and effort into developing your presentation, that is, the pitch deck. Getting advice and assistance from a professional advisor may be well worth the cost. The pitch is crucial to win the investor's confidence.
- Research, identify, and target the appropriate investor(s). The wise inventor will often hire a fund raiser with direct experience in the field of interest.

Chapter 9

An Introduction to the National Institutes of Health SBIR/STTR Programs

Deepa Narayanan and Michael Weingarten
SBIR Development Center, National Cancer Institute, Rockville, MD, United States

OVERVIEW AND GOALS OF THE PROGRAM

The Small Business Innovation Research (SBIR) and Small Business Technology Transfer (STTR) programs are one of the largest sources of early-stage funding for start-up companies in the field of life sciences. SBIR and STTR are congressionally mandated programs. Federal agencies with extramural research budgets over $100 million are required to set-aside a certain percentage of their budget to SBIR, and those with research budgets over $1 billion are required to set aside a portion of these funds for STTR to support small businesses to develop innovative technologies. By supporting innovative life-science projects through preclinical and clinical development, the ultimate goal of the NIH is to help commercialize new biomedical technologies, thereby improving public health. In 2016, the total available funding for the SBIR and STTR programs was $982 million across 27 different Institutes of the NIH.

Key Differences Between SBIR and STTR

When deciding to apply for an SBIR or STTR award, it is important to understand the differences between the two programs. The amount of funding available to SBIR awards is much greater than that available for STTR. This funding is based on congressional legislation which sets aside currently 3.2% of the NIH extramural budget for SBIR and 0.45% for STTR. Under the SBIR program, companies may partner with US research institutions such as universities, but it is not a requirement. In addition, under SBIR, companies can outsource 33% of their Phase I activities and 50% of their Phase II activities. To be eligible for SBIR, the primary employment of the principal investigator (PI) (i.e., >50%) must be with the small business for the period of the project. The STTR

Medical Innovation. https://doi.org/10.1016/B978-0-12-814926-3.00009-7

program, meanwhile, has different requirements. The primary goal of the STTR program is to facilitate the transfer, development, and commercialization of a technology conceived in academia to a small business. As such, STTR requires a small business to partner with a US research institution. One of the benefits of applying for the STTR program is that it does not mandate that the PI is employed primarily by the small business. This important clause allows faculty researchers to stay at the university and continue commercializing their research. In terms of funding allocation for STTR projects, a minimum of 40% of the work should be conducted by the small business, and at least 30% should be conducted by the research institution. In addition, there must be an intellectual property (IP) agreement between the research institution and the small business, providing to the company IP rights necessary for carrying out follow-on R&D and commercialization. It is critical, however, to note in both SBIR and STTR, the awards are always made to the small business.

Eligibility

A company is eligible to apply for SBIR or STTR funding if it meets the following criteria:

1. The company must be a small business concern (SBC) [1].
2. The company is a United States for-profit business (based in the United States and work performed in the United States).
3. The company has 500 or fewer employees including affiliates.
4. The company is >50% owned by US individuals and independently operated.

OR

5. The company is >50% US owned and controlled by another one business concern that is >50% US owned and controlled by one or more individuals.

OR

6. The company is >50% US owned by multiple venture capital operating companies, hedge funds, private equity firms, or any combination of these (for SBIR only).

Program Structure: Funding Phases

The SBIR and STTR programs consist of three phases (Fig. 9.1). Companies can apply initially for a Phase I SBIR or STTR award from the NIH. Under a Phase I, funding caps differ by Institute at the NIH but generally range from $225,000 to $300,000 in total costs for a period of 6 months to a year. The goal of a Phase I is to conduct a proof of concept or feasibility study. If the project is successful, then a small business can apply for a Phase II award, which

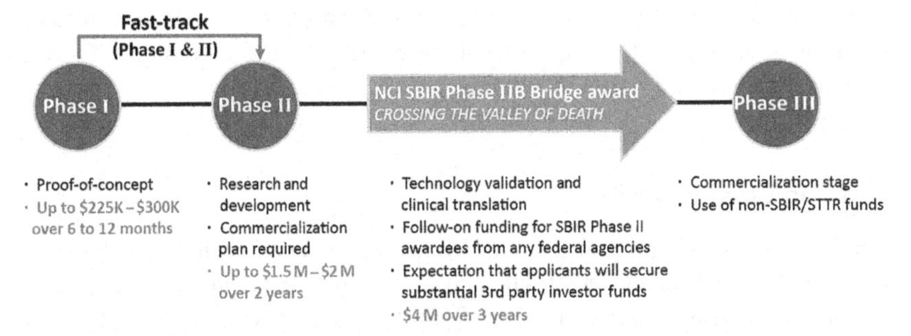

FIG. 9.1 SBIR/STTR programs: a three-phase program.

involves the full R&D of the technology. Funding caps for Phase II also differ by Institute, but generally companies can apply for a range of $1 million to $2 million for a period of 2 years. Companies that already have substantial preliminary data also have the option of applying for an SBIR or STTR Fast-Track award. A Fast-Track award combines Phase I and II applications, allowing the applicant to go through the NIH peer-review process just once as opposed to twice. Companies should speak first with an NIH program director before applying for a Fast-Track to make sure they are a good fit for this mechanism. In 2009, the National Cancer Institute (NCI) recognized that many life-sciences companies face a funding gap, commonly known as the "Valley of Death," that often exists between the end of the SBIR Phase II award and the next round of nonfederal financing needed to advance a promising technology toward commercialization. To bridge this gap, NCI designed a special funding program called the Phase IIB Bridge award to incentivize partnerships between federally funded SBIR Phase II awardees and third-party investors and/or strategic partners. These partnerships are critical to the long-term viability of these SBIR Phase II projects and their ultimate success in reaching the cancer community, especially patients. Under the Phase IIB Bridge Program, NCI provides up to $4 million over 3 years for further development and validation of the technology. To be competitive for the Bridge funding, applicants are expected to secure third party, nonfederal funds that equal or exceed the requested NCI funds.

Over the past 9 years, the Phase IIB Bridge program has achieved its goals. The NCI has funded cumulatively $51 million in projects that have raised just over $220 million from private investors during the grant award periods—almost a 4:1 match. Companies that have completed their Bridge awards have raised an additional $167 million in cumulative, follow-on private investment to enable continued development of Bridge award-funded technologies. A number of these projects have since been commercialized. Other NIH institutes, notably the National Heart Lung and Blood Institute (NHLBI) and the National Institute of Neurological Disorders and Stroke (NINDS) also offer Phase IIB funding now. It is important to review specific funding information at target institutes when planning an application, because each Institute at the NIH sets its own funding caps for each stage of funding.

Funding Opportunities

The SBIR and STTR programs at the NIH are managed independently by the 23 individual Institutes and Centers (ICs) that participate in the program (Fig. 9.2). Management models for SBIR and STTR Programs vary by Institute [2], but each IC has identified specific scientific and research points-of-contact [3]. Applicants are encouraged to speak with the appropriate SBIR and STTR program manager in that IC before submitting their application. The most popular funding opportunity is the NIH SBIR/STTR Omnibus Solicitation, published annually with three receipt dates each year. The Omnibus Solicitation is an NIH-wide funding opportunity announcement that accepts all investigator-initiated applications that fit the mission of the NIH. In addition, the NIH publishes one annual SBIR contract solicitation that invites proposals from companies working on clearly defined topic areas that are considered priority areas for each IC. In addition, there are a number of targeted grant solicitations released by various ICs that focus on specific research areas and/or institute priorities. Application due dates, budgets, and topic requirements vary for each solicitation, but they are listed and updated periodically on the NIH SBIR and STTR website [4].

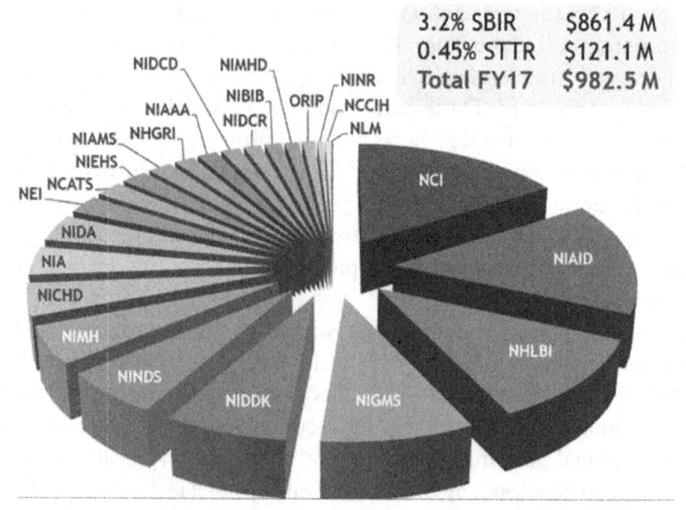

FIG. 9.2 NIH SBIR/STTR budget allocations for FY2017. There are 23 NIH institutes participating in the NIH SBIR/STTR program. Each institute has a different budget based on the extramural budget of the IC. NCI has the largest SBIR/STTR budget followed by NIAID and NHLBI.

BUILDING A SUCCESSFUL APPLICATION

The SBIR and STTR programs are often the first source of funding for entrepreneurs who are propelling technologic innovation. These programs are quite competitive across the NIH. In 2016, 12.6% of all Phase I SBIR applicants received an award, and 25% received Phase II SBIR awards [5]. For STTR awards, the funding success rates were 14.2% and 31.4% for Phase I and Phase II, respectively. From 2007 to 2016, the NIH saw a 60% increase in the total number of SBIR and STTR applications received, and correspondingly, the overall funding success rate decreased from 21.2% to 15.3%. These rates vary for both SBIR and STTR programs each year depending on the availability of funds for new competing grants, as well as the number and quality of applications received by NIH. Due to the competitive nature of the programs, it is important for applicants to understand the NIH funding process before applying [6]. Unlike basic research grants that prioritize hypothesis-driven research and advancement of scientific knowledge (discovery), the SBIR and STTR programs tend to be product-focused and favor those applications with a strong potential for commercial development. In the next section, we offer some tips to help put together a successful application.

Application Tip 1. Start Early

Navigating the administrative requirements of federal grants can be challenging for new applicants. The SBIR and STTR application process infographic [7] (Fig. 9.3) has useful information particularly for first-time applicants. Currently, the NIH requires that companies prior to the application process have an Employer Identification Number and a Dun and Bradstreet Universal Numbering System (DUNS) number and register at the System for Award

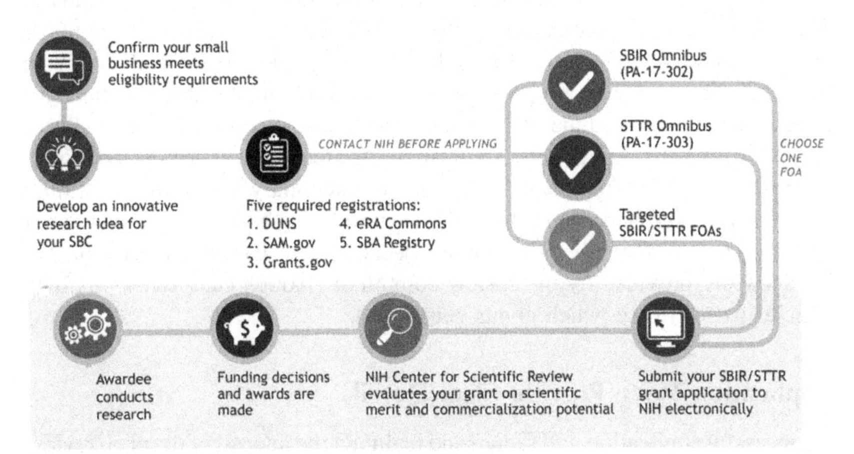

FIG. 9.3 NIH SBIR/STTR program application flowchart.

Management (SAM) and the SBA Company Registry. Other requirements include an account with the NIH Electronic Research Administration system and at Grants.gov—a portal for applicants to find and apply for federal grant funding. Prior to applying, it is recommended that companies identify the NIH Institute and appropriate funding opportunity announcement that best fit their technology. In some cases, the technology may be cross-cutting and might be appropriate for more than one IC. Starting the application process early helps the applicant to refine the vision of their product by initiating discussions with technical experts, potential customers, payers, investors, commercialization partners, and other stakeholders.

Another critical task is assembling a strong scientific and commercial team, including a PI with the right expertise to lead the project. It is very rare that a small business would have all the necessary expertise and capabilities needed to complete a successful SBIR grant. Applicants should consider partnering with collaborators, consultants, or Contract Research Organizations (CRO) to fill in any gaps. Companies without a track record or expertise in commercializing biomedical technologies should also consider finding a partner with business experience. Applicants are allowed and encouraged to contact NIH Program Officers to discuss their concepts and receive feedback prior to submitting a grant application.

Application Tip 2. Understand What Peer Reviewers Want

Most grant applications, unless they are targeted solicitations requiring special review, are received and reviewed at the NIH Center for Scientific Review (CSR). The Receipt and Referral Office at CSR reviews each application and assigns it to an IC and review panel. It is important for applicants to recognize that applications are peer-reviewed by non-NIH scientists with expertise in the applicant's research area and by business professionals with backgrounds in commercialization of similar technologies. Therefore, a well-written application should address strategies for both the science and commercialization. In terms of the science, a proposal must have sufficient technical details for reviewers to understand the approach. The five review criteria used to evaluate the application are Significance, Innovation, Approach, Investigators, and Environment (Fig. 9.4). Applicants will receive a summary statement containing detailed feedback with strengths and weaknesses for each of the five criteria within 4–6 weeks of review. Funding decisions are based primarily on the impact score provided by the review committee, though each IC has its own policies to determine which grants get funded.

Application Tip 3. Provide Clear Details

A successful application will secure and maintain the interest of the peer-review committee. The Specific Aims page is the most critical part of an application

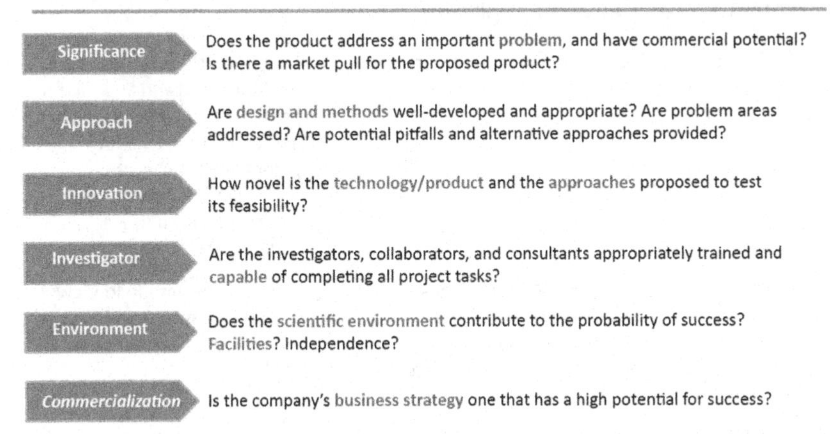

FIG. 9.4 Review criteria for SBIR/STTR programs.

and should be treated as a standalone page from which a reviewer can gain a reasonable understanding of critical components of the project without reading other parts of the application. Applicants are required to limit their Specific Aims to one page. The page must demonstrate a clear product description, provide an explanation on the innovation of the product and how it is different from the current state of the art, and convey the unmet clinical need the technology aims to address. The Specific Aims page should also state the technical objectives of the project, accompanied by appropriate milestones and/or clearly articulated success criteria. Whenever reasonable, the success criteria should be defined by quantitative metrics. In cases where only qualitative success criteria are appropriate, they should be stated clearly. For Fast-Track applications, the applicant should present criteria for a go or no-go decision at the end of the Phase I component.

The research strategy section must provide a detailed technical plan for achieving the specific aims, describe potential pitfalls, and include alternative approaches. The proposed work must be within the scope of time and budget requirements of the funding solicitation. While preliminary data are not required for a Phase I application, competitive applications do provide prior experimental data and/or a clear scientific rationale for their proposals. Letters of support from collaborators and/or key opinion leaders are often powerful endorsements of the technology for peer review. The commercialization plan is a key component of a Phase II application and must include: a description of the value and impact of the product; a market, customer, and competitive landscape analysis; a description of the regulatory and/or reimbursement strategy; the company's finance, production, and marketing plans; and a description of the company's plans for revenue generation. Other application components, such as bio-sketches for PI and key personnel, a detailed budget for each project

TABLE 9.1 Useful Tools for Applicants

NIH reporting and tracking tool	https://projectreporter.nih.gov/	NIH Reporter provides access to reports data and analysis for previously funded NIH grants and could be useful for potential applicants to understand competitive landscape and find collaborators
CSR receipt, referral, and review process	https://public.csr.nih.gov/ApplicantResources/ReceiptReferal/Pages/Submission-and-Assignment-Process.aspx	Center for Scientific Review (CSR) process for receipt, referral, and assignment of applications
SF424 guide	https://grants.nih.gov/grants/how-to-apply-application-guide/forms-d/sbir-sttr-forms-d.pdf	SF424 SBIR/STTR Application Guide for NIH and other PHS Agencies
Application guide	https://grants.nih.gov/grants/how-to-apply-application-guide.html	Application guide to submit grant applications to NIH and other Public Health Service Agencies

period including those for subcontractors, and information on facilities and equipment, should be provided in the appropriate NIH format [8]. NIH offers several useful websites and tools to help companies put-together successful applications (Table 9.1).

Application Tip 4. Remember the Postsubmission Requirements

It can take 6–9 months for an application to get funded by the NIH. Unsuccessful applicants will receive feedback from peer review in the form of a summary statement and can contact the assigned Program Officer to discuss the feedback and strategize about resubmitting the applications. Resubmissions typically have a greater funding rate than first-time applicants, especially if the applicant has addressed adequately the concerns in the summary statement. For both funded and non-funded applicants, the Program Officer serves as the main point-of-contact at the NIH. If funded, companies must provide periodic technical reports to the NIH about their progress. Awardees are encouraged to maintain regular contact with their Program Officers and inform them of technical or commercialization milestones achieved by the company. The Program Officer can also serve as a mentor and guide for the applicant toward other funding and non-funding resources (both federal and non-federal) that can help accelerate the path to commercialization.

SUCCESS STORIES FROM THE NIH SBIR AND STTR PROGRAMS

Since its inception in 1982, the NIH SBIR program has provided the small business community with critical seed funding to support the development of a broad array of commercial products for the detection, diagnosis, treatment, and prevention of disease. In 2015, The National Academy of Sciences (NAS) evaluated the performance of the NIH SBIR and STTR programs against the broad congressional objectives for the SBIR and STTR programs and found that the program has an overall positive impact [9]. Of the 570 SBIR/STTR respondents to the NAS survey, 97% reported that the funding had a positive effect on the recipient company, while 62% reported a transformative effect. One key metric of success for the SBIR program at the NIH is successful product commercialization. The 2015 survey of funded respondents showed that 49% of SBIR and STTR respondents had successfully achieved either sales or licensing revenues at the time of the survey, and a further 25% expected sales in the future. In addition, more than 80% of respondents indicated that their project received additional external investment in the technology related to the surveyed project; 74% of the respondents remarked that the continuation of the project would have been difficult without SBIR or STTR funding. By providing funding opportunities and other key resources for these early-stage companies, SBIR and STTR programs facilitate the translation of innovative life-science technologies to products or services that can improve public health.

Many companies [10] have developed or validated successful life-science products with funding from the NIH SBIR and STTR programs. A notable success story is Illumina, Inc., now a leader in DNA sequencing technologies. As a start-up, Illumina received NIH SBIR funding between 1999 and 2004 to contribute to the development of its genotyping, parallel arrays, and gene expression profiling technologies—all of which played important roles in Illumina's growth. Dr. Mark Chee, the founder of Illumina, has pointed out that the funding from the NIH SBIR program propelled the development of the company's core technologies at a very early stage, when obtaining funding from private investors would have been difficult [9]. Today, the company generates over $2 billion a year in sales. Similarly, TomoTherapy received SBIR awards to validate key enabling factors for its integrated image-guided technology for radiation therapy. While the technology was developed with academic grants, it was the SBIR awards that propelled the commercialization of their advanced image-guided radiation therapy technology. After a successful 510(k) clearance in 2002, the company went public in 2007 and was ultimately acquired by Accuray for $277 million in 2011 [11].

More recently, Lift Labs found itself in the media spotlight after Google announced its acquisition of the NIH-funded small business [12]. Lift Labs developed a spoon that uses sensors to detect hand tremors and counteract them using Active Cancellation of Tremor technology to minimize food spillage for patients with essential tremor or limited hand mobility due to neurodegenerative diseases such as Parkinson's. The company utilized SBIR funding from the NIH to continue the development of this technology through clinical research and additional R&D. Another recent commercial success of the NIH SBIR and STTR programs was a minimally invasive technology for lung cancer screening that was initially developed using academic funding. The technology was spun-off from Boston University as Allegro Diagnostics and received SBIR funding to further develop the technology. The technology is a first-of-its-kind bronchial test for lung cancer that allows for an early diagnosis of the disease by scanning and detecting genomic alterations often found in people with lung cancer. Allegro was acquired by Veracyte for $21 million in 2014 [13], and the latter launched the technology, now titled Percepta, in the US market in 2017.

The success of these ventures illustrates the role of SBIR and STTR programs in incentivizing and supporting the commercialization of innovative medical technologies that might otherwise struggle to advance due to difficulties in obtaining early-stage funding.

SBIR AND STTR PROGRAMS: MORE THAN JUST FUNDING

The SBIR and STTR programs are evolving and adapting continuously to fulfill the needs of the small business community as the medical innovation climate changes. While SBIR and STTR funding is critical for R&D, there are a wide range of additional challenges that companies face as they grow and advance their technologies into the clinic. The NIH offers programs and resources to help companies overcome these challenges. Some of these programs are led by the NIH and some are managed by different ICs (Fig. 9.5).

NIH Technical Assistance Programs: The NIH offers two technical assistance programs for awardees each year. The Niche Assessment Program (NAP) is open to Phase I awardees and is intended to provide market analysis to help small businesses develop a viable commercialization plan—a critical component of the Phase II application. The analysis is provided by an external consultant group, Foresight, with insight into the market size, competing technologies, and potential barriers to entry to help companies strategically position their technologies in the market. The Commercialization Assistance Program (CAP) is also a NIH-managed 9-month program that has been offered since 2004 to Phase II and Phase IIB awardees across the NIH, the Centers for Disease Control (CDC), and the Food and Drug Administration (FDA). The program is aimed at assisting participants with evaluating their commercialization options based on their specific technologies and specific needs. These resources are complimentary but do require a separate application process.

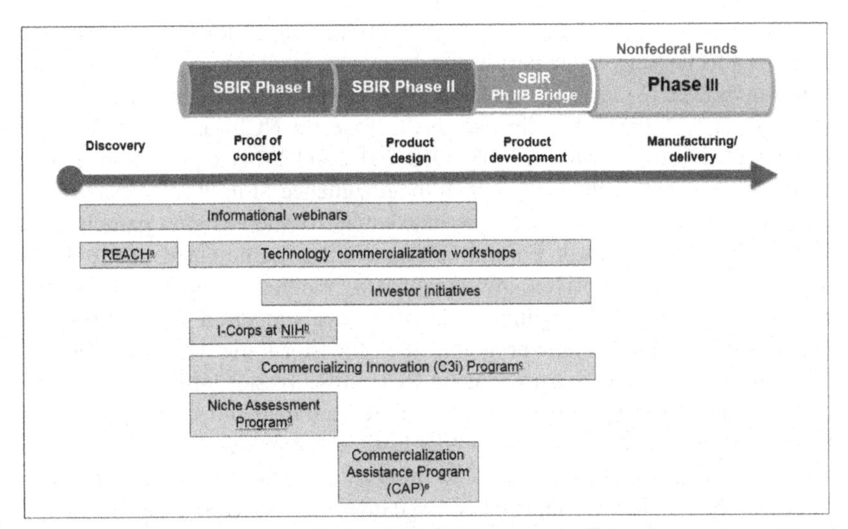

FIG. 9.5 Additional resources available to SBIR/STTR community. [a]https://grants.nih.gov/grants/guide/rfa-files/RFA-OD-14-005.html. [b]https://sbir.cancer.gov/programseducation/icorps. [c]https://grants.nih.gov/grants/guide/pa-files/PA-17-286.html. [d]https://sbir.nih.gov/nap. [e]https://sbir.nih.gov/cap.

Innovation Corps (I-CORPS at NIH): The I-Corps at NIH program is an entrepreneurial training program managed by the National Cancer Institute (NCI) and is available to Phase I SBIR and STTR awardees from 16 ICs across the NIH and CDC. Based on the highly successful NSF I-Corps program, the I-Corps at the NIH curriculum is designed to provide life-science researchers with real-world, hands-on entrepreneurship training, facilitated by domain experts from the biotech sector. The I-Corps at NIH program has been modified to focus on the life-sciences sector and is designed to support training that will help NIH- and CDC-funded small businesses overcome key obstacles along the path of innovation and commercialization. Many NIH-funded start-ups are run by founders who have spun their technologies out of a university. While these founders are outstanding scientists, many lack business skills important for managing their small businesses. I-Corps at the NIH teaches start-ups how to build a strong business model best suited for the technology they are developing. The program is focused on experiential learning where researchers are required to "get out of the building" and talk to potential customers and key stakeholders. Feedback from these interviews informs the "Business Model Canvas," which provides a framework for companies to define their value proposition, customer segments, partnership strategy, and financing needs, as well as to begin to address their regulatory and reimbursement strategy. As of the end of 2017, 100 teams have successfully completed the program, and over 90% of surveyed graduates rated the program "very good" or "excellent."

Connecting companies to investors: Federal funding alone is not sufficient to move a technology from early-stage development all the way into the clinic. Companies need to raise external funds, especially if they are in the field of drug or device development. To help address the issue, the NCI was the first institute to conduct investor forums (in 2009, 2010, 2012, and 2014) to provide an opportunity for selected companies to pitch to an audience of invited investors, and this program has been successful; as an example, 16 months after participation in the 2014 NCI Investor Forum, companies that participated in the event reported 25 deals with private investors that were either in negotiations, ongoing, or completed. NCI facilitated half of the deals reported. Since 2014, the NCI has implemented a new program called Investor Initiatives. SBIR-funded companies are selected to participate in NCI SBIR Investor Initiatives based on an assessment from a group of over 50 venture capitalists and strategic partners. Currently the NCI, along with other NIH Institutes, provides funding assistance for selected companies to pitch their technology at leading private investor conferences, such as the redefining early-stage investments (RESI), the BIO Investor Forum, and the Life Sciences Summit. In 2016, 22 NCI SBIR awardee companies were selected to present at seven separate events, and participants reported over 260 meetings with potential investors or strategic partners. Other institutes have joined NCI in reaching out to investors, providing pitch coaching and mentoring assistance to selected companies.

Workshops and webinars: Several ICs conduct workshops for awardees or potential applicants to provide specific guidance designed to help with the application preparation or commercialization process. NHLBI hosts periodic webinars with regulatory and other domain experts to provide guidance or training to help awardees advance their research. The NCI's SBIR Development Center hosts a specially-designed workshop to provide awardees an opportunity to learn how to utilize federal and local resources for advancing their cancer technologies to commercialization. Some workshops are institute-specific, while others are in collaboration with other ICs and include representatives from multiple institutes.

Outreach events: The NIH has a very active outreach program that is typically managed individually by each IC. During outreach events, Program Officers provide information on upcoming funding opportunities and are also available to meet one-on-one with prospective applicants to discuss their project idea. These outreach programs are designed to encourage participation in the program by promising start-ups. Targeted outreach efforts are also directed toward socially and economically disadvantaged small businesses (SDB) and women-owned small businesses (WOSB) to increase their participation [14] in the program. The NIH also hosts the Annual National SBIR/STTR Conference targeted for new small businesses, university researchers looking to spin off companies, and prospective investors interested in NIH-funded companies with the goal of providing an overview of and opportunities available from the programs at NIH. Typically, Program Officers from all NIH ICs, in addition to CDC and FDA, attend the conference each year.

SUMMARY

Medical innovation is driven primarily by small businesses which are affected disproportionately by the funding environment. The NIH SBIR and STTR Programs have proven to be a vital source of support for small business entrepreneurs in the United States by providing stable, early-stage funding for companies developing innovative life-science technologies aimed at improving public health. The NIH SBIR and STTR programs are constantly adapting to the ever-changing landscape of medical innovation. It is important to realize that in addition to grant and contract funding, the NIH provides mentoring and other assistance to ensure that technologies reach patients. The NIH SBIR and STTR programs seek to support exciting new technologies from innovators that can improve patients' health and meet the NIH mission of applying scientific knowledge to enhance health, increase both the duration and quality of life, and decrease illness and disability. The NIH encourages new applicants to introduce their technologies to Program Officers to discuss potential funding strategies and seek guidance.

IMPORTANT POINTS

- The NIH SBIR and STTR Programs allow US owned and operated small businesses to engage in federally funded research to develop and commercialize innovative technologies to improve health and save lives. The primary difference between SBIR and STTR programs are that while the SBIR program permits partnering with non-profit research institutions such as universities, the STTR requires a formal partnership.
- Developing a competitive application is critical to get SBIR/STTR funding. Applicants are strongly encouraged to contact NIH program staff prior to application to discuss their technology and specific aims.
- SBIR and STTR programs at the NIH provide more than just funding. They offer awardees access to a number of useful resources and programs to accelerate the commercialization of SBIR-funded technologies and to help build your network of investors, peers, and collaborators.
- NIH SBIR programs are managed independently by each institute/center and vary in terms of the funding opportunities and resources provided to applicants and awardees. The I-Corps at NIH program is an entrepreneurial training program managed by the NCI for the NIH and is designed to provide life-science researchers with real-world, hands-on entrepreneurship training, facilitated by domain experts from the biotech sector.

ACKNOWLEDGMENTS

The authors would like to thank M. Lisa Yeom for writing assistance, technical editing, language editing, and proofreading.

DISCLOSURES

None.

REFERENCES

[1] Qualifying as a small business, https://www.sba.gov/contracting/getting-started-contractor/qualifying-small-business. Accessed 23 October 2017.

[2] National Academies of Sciences, Engineering, and Medicine. SBIR/STTR at the National Institutes of Health. Washington, DC: The National Academies Press; 2015. Program Initiatives at NIH. https://doi.org/10.17226/21811.

[3] HHS SBIR/STTR Agency. Contact Information, https://sbir.nih.gov/engage/ic-contacts; 2017. Accessed 16 October 2017.

[4] SBIR/STTR Funding, https://sbir.nih.gov/funding; 2017. Accessed 16 October 2017.

[5] NIH SBIR/STTR Award Data, https://sbir.nih.gov/statistics/award-data; 2017. Accessed 16 October 2017.

[6] Beylin D, Chrisman CJ, Weingarten M. Granting you success. Nat Biotechnol 2011; 29(7):567–70.

[7] SBIR/STTR Application Process infographic, https://sbir.nih.gov/infographic; 2017. Accessed 16 October 2017.

[8] General instructions for NIH and other PHS agencies, https://grants.nih.gov/grants/how-to-apply-application-guide/forms-d/general-forms-d.pdf; 2017. Accessed 16 October 2017.

[9] National Academies of Sciences, Engineering, and Medicine. SBIR/STTR at the National Institutes of Health. Washington, DC: The National Academies Press; 2015. Quantitative Outcomes. https://doi.org/10.17226/21811.

[10] Rao GN, Williams JR, Walsh M, Moore J. America's seed fund: how the SBIR/STTR programs help enable catalytic growth and technological advances. Technol Innov 2017; 18(4):315–8.

[11] Accuray to acquire TomoTherapy [Press Release], http://de.accuray.com/pressroom/press-releases/accuray-acquire-tomotherapy; 2011. Accessed 16 October 2017.

[12] News article by Laura Lorenzetti, Google buying Lift Labs, the maker of a self-stabilizing spoon, http://fortune.com/2014/09/10/google-buys-lift-labs-the-maker-of-a-self-stabilizing-spoon/, 2017, Accessed 16 October 2017.

[13] Veracyte, Inc. to Acquire Allegro Diagnostics. Accelerating entry into pulmonology market, http://investor.veracyte.com/releasedetail.cfm?releaseid=869242. Accessed 16 October 2017.

[14] National Academies of Sciences, Engineering, and Medicine. Innovation, Diversity, and the SBIR/STTR Programs: Summary of a Workshop. Washington, DC: The National Academies Press; 2015. https://doi.org/10.17226/21738.

Chapter 10

Clinical Trials for Medical Device Innovators

William D. Voorhees III* and Theodore W. Heise†
*MED Institute Inc., West Lafayette, IN, United States, †Cook Incorporated, Bloomington, IN, United States

DESIGNING CLINICAL TRIALS

Perhaps the most exciting and rewarding time in the development of a new medical device is the first time one sees it work successfully in a patient, but the ensuing clinical trial can be the most exhausting, frustrating, expensive, and time-consuming part of the entire enterprise. So before rushing headlong into the unknown, consider the following: "Is it necessary for me to conduct a clinical trial of my new device?" The answer is "It depends."

It depends on what you are trying to accomplish. Maybe you just want to show that your device works (i.e., the Proof of Concept). Perhaps you have reached a point at which no more can be learned without testing in humans, for example, pain relief cannot be studied on the bench at all and rarely well in animals. Your goal may be to demonstrate product efficacy to investors and to raise additional funds. Maybe all you want is enough clinical data to improve your chances of selling the rights to the intellectual property. Perhaps you just want to enhance market awareness and exposure to the device (such as having a few key opinion leaders use it and write a paper), or maybe you want to take it all the way to regulatory approval with proof of safety and effectiveness, which is collected typically in what is called a pivotal study. Even if you want to own the regulatory approval of your device, keep in mind that a clinical study may not be needed if you can show that all relevant risks can be mitigated by bench and/or animal testing or with appropriate labeling. This is why a rigorous risk-analysis is imperative to your plan for development of the device.

Assume you have developed your device to the point that you have made prototypes that you have bench-tested adequately to satisfy yourself that they function as expected and they appear safe to use in humans. The "Grandmother Test" is a good benchmark; would you let your grandmother be treated with this device? Even if the answer is "Yes," you probably need a clinical feasibility

Medical Innovation. https://doi.org/10.1016/B978-0-12-814926-3.00010-3

study first. Conducting a clinical feasibility study may also be a good first step when only formal testing in humans can provide the evidence that the device offers the expected benefit (Proof of Concept) or is required for making any final changes in design (e.g., to apply to specific human anatomy). Such data can simply be gathered no other way than to test in humans. Even if you know you will need a pivotal study for regulatory approval, it is likely that you will need initial information about how large a therapeutic effect the device will have before you can design a larger pivotal study to compare its results with state-of-the-art therapy (often a currently marketed device) in a statistically rigorous clinical study. A feasibility clinical study can be used to provide this crucial information.

If your interest lies only in developing the new device to the point that the concept (intellectual property) can be sold to another company to finalize its development and take it to commercialization, then conducting a clinical feasibility study may be adequate to provide the information that the company planning to purchase your device will need to decide to close the deal.

REGULATORY CONSIDERATIONS FOR CLINICAL STUDIES

For regulatory purposes, clinical trials are grouped into two basic varieties: significant risk (SR) studies and nonsignificant risk (NSR) studies. The Food and Drug Administration (FDA) prefers that IRBs (Institutional Review Boards) make this determination; however, many IRBs put this back on the FDA to decide. Caution: If your study will involve multiple investigative sites and multiple IRBs, if any one of the IRBs decides it is a SR study, then the whole study will need to be so designated. You may wish to be proactive and write to the chief of the branch of the FDA which will hold authority over your device, describing your study and requesting determination by the FDA regarding the risk status of the study. This decision would then govern how your study is conducted.

All SR studies require an IDE (investigational device exemption). If your device is considered to present SR to patients, you will need to apply for an IDE and obtain FDA approval prior to conducting the study. In contrast, NSR studies only require approval of an IRB. If an IDE is required, an IDE Early Feasibility Study now may offer some leeway with respect to the amount of nonclinical testing required to support the IDE, so it is likely worthwhile to try this route. The alternative is a traditional IDE which often requires all nonclinical testing (bench and animal) to be completed prior to applying for an IDE.

If you conduct a clinical feasibility study and decide that you want to complete development of the device and obtain regulatory approval for it on your own, this may be the perfect time to meet with the appropriate authorities of the FDA to discuss your plans and obtain their advice. Assuming a pivotal clinical study will be necessary to obtain data for approval, early in the planning stage of the study is a good time for a "presubmission" meeting (in this case, probably a "pre-IDE" meeting). Your request for the meeting will include a description of the device, proposed indications for use, overview of product development, planned nonclinical testing, the design for the proposed clinical study, and specific questions for the FDA. The FDA makes available numerous guidance documents on their processes.[1] In this meeting, you can determine what nonclinical testing the FDA will require in order to approve the IDE, the endpoints and follow-up for the clinical study, and agreement that the design of your proposed clinical protocol is likely to prove adequate to yield sufficient analyzable data for determining if the device is approvable. FDA reviewers/scientists generally look for objective, measureable endpoints (e.g., procedural success, long-term success, rates of adverse events). If you are also interested in collecting data to support a coverage decision, third-party payers tend to be interested in more subjective endpoints (e.g., quality of life and effect on patients), which may be difficult to validate. Nevertheless, it is beneficial to work these subjective endpoints into the pivotal study so that the data are available when it is time to discuss cost coverage for your device.

The most important thing you can bring to a pre-IDE meeting is a well-reasoned clinical protocol. Here is where decisions begin to become difficult. What kind of a clinical trial is needed? Your options for the design of a clinical study may be broader than you think. Investing in consultation with an experienced biostatistician may be well worth the time and investment at this point. Selecting the trial design that maximizes the likelihood of obtaining relevant data while minimizing time and expense is both a science and an art; therefore, do your homework and choose a statistician you trust.

Your options for a pivotal clinical trial will depend to some extent on what is already known about your device and the condition it will be used to treat; is it an incremental improvement on a current product, or is it a novel technology? Do you plan to show superiority to the current standard of care with respect to effectiveness, or do you expect to show noninferiority, while perhaps demonstrating an improved safety profile or a less expensive device?

Of course, the gold standard for clinical trials that always comes to mind first is the prospective, blinded, randomized, controlled study. Such studies can be large and expensive, but if recent, well-conducted studies have been published using the standard of care, you may very well be able to use their results to develop a performance goal to which your new device can be compared given sufficient data from an observational study with a single-arm registry and appropriate hypothesis testing. Having an experienced biostatistician is crucial to developing the most efficient study design possible and justifying it to the appropriate regulatory authorities.

1. https://www.fda.gov/downloads/MedicalDevices/DeviceRegulationandGuidance/GuidanceDocuments/UCM311176.pdf

MANAGING CLINICAL TRIALS

Except for small, clinical feasibility studies, unless you are an experienced clinical trialist, do not try to manage a clinical trial on your own. The larger the trial, the more help you will need. Hiring a contract research organization (CRO) to manage the trial may be your best, or only, option, but be prepared for the expense. Similarly, if FDA submissions are needed, you may be wise to seek the help of seasoned regulatory affairs professionals.

Conducting a clinical trial entails a lot of individual tasks. Your study design will have determined the number of patients you need to enroll. How many investigative sites will you need? This decision involves a trade-off between cost and speed of enrollment. Each site costs on the order of $75,000 (2017 USD) to bring up. You may wish to set a minimum number of patients to be enrolled per site, but that is no guarantee that each site will come through for you. Select sites judiciously, which may require on-site screening visits. Be careful, because the inclusion of a site that enrolls no, or only 1 or 2, patients is expensive and demoralizing. Remember, you will need a contract with each site, and each site will need local IRB (Institutional Review Board) review and approval unless you have contracted with a central IRB to oversee the entire study. Currently each IRB submission also usually costs money. You will need to perform an initial site visit, often with some level of training on the protocol and data entry for the investigators and the personnel who will actually be doing the study and collecting the data. Interim monitoring visits are required to ensure compliance with the protocol and to compare source data with data entered in the database according to a predetermined monitoring plan. Closeout visits are also needed at the end of a trial. As data come in, it is rarely perfect. Someone will be needed to review incoming data and prepare queries to the sites as necessary; we cannot stress how important this activity is to ensure the integrity and quality of the data.

Larger, more complex, multicenter clinical trials may also require (or benefit from) one or more of the following management tools.

Clinical Events Committee (CEC)

A group of unbiased physicians knowledgeable in the field of interest, typically charged with adjudicating whether or not individual serious adverse events (SAEs) are related to the use of the device or to the procedure under study.

Data Safety Monitoring Board (DSMB)

A group of unbiased physicians not involved in the trial but knowledgeable in the area. They are charged with developing rules for halting the trial based on anticipated aggregate rates of SAEs.

Core Laboratory

A central laboratory set up to provide unbiased analysis of specific results (e.g., imaging, measurements, or analysis of all clinical specimens); such a laboratory provides consistency in the analyses.

Smaller studies, particularly early feasibility trials, will need this level of oversight only rarely.

ANALYZING THE RESULTS

Once again, smaller trials or observational studies that are simply reporting means and standard deviations and perhaps simple *t*-tests for comparisons may not require an independent biostatistician. In contrast, more complicated studies requiring complicated statistical analyses, studies with multiple endpoints, and studies requiring computational modeling, are all likely to need a biostatistician with particular expertise in the analysis of complex study designs.

COST DRIVERS FOR CLINICAL STUDIES

There is no denying that clinical trials are expensive propositions. There is little economy of scale; by and large, big pivotal trials cost proportionally more than small studies. The best thing one can do is design the study well to collect only the most relevant data, size it properly based on solid statistical grounds, select investigative sites judiciously to provide sufficient numbers of subjects, and manage the trial efficiently. These actions will not make a study inexpensive, but emphasis on consistency and paying strict attention to all the details will help minimize the cost.

The cost drivers for clinical studies are numerous, vary between studies, and are certainly subject to change over time. Rather than attempt to attach a current monetary value to each cost driver, it may be as valuable to provide a list of these drivers so that you, as the study sponsor, understand what estimates you will need to price the study. The following is a list of items that must be considered when estimating the cost of a clinical trial:

- Starting up one investigative site:
  - contract costs
  - IRB review (at each site or single central IRB?)
  - site screening visit (if necessary)
  - site initiation visit
  - initial device inventory
- Data management:
  - creation of a database
  - creation of a data collection system [typically electronic data collection (EDC)]
  - training site personnel on data entry

- Data collection:
  - reimbursements to site for completed data forms (typically at patient screening, procedure, and follow-up; priced per form completed)
  - reimbursement to site for tests and procedures that are not "standard of care" for the disease entity or procedure
  - compensation to patients for travel and parking if nontrivial
  - data review and queries
- Project management:
  - monitoring visits (interim visits [how many?], final closeout visit)
  - establish or contract with a Clinical Events Committee (if needed)
  - how many meetings over course of study?
  - establish or contract with a Data Safety Monitoring Board (if needed)
  - how many meetings over course of study?
  - contract with core laboratories, for example, imaging (if needed)
  - final data analysis (statistician) and preparation of the manuscript

Regulatory costs for conducting clinical studies are minimal. You may have the cost of a meeting with FDA if you decide to have a pre-IDE meeting; however, these meetings can be held by teleconference. There is no fee to submit the IDE application itself. The primary regulatory costs come when applying for market approval. If your device requires a 510(k) application, the fee is $4690 (USD 2017). For a premarket approval application (PMA) for a Class III device, the fee is $234,495 (USD 2017). If you qualify as a small business, the 2017 application fees for a 510(k) and PMA are $2345 and $58,624, respectively. Small businesses with an approved small business designation are eligible to have the fee waived on their first PMA.

Clinical trials are exciting, exhausting, and expensive. As an innovator of a new medical device, you need to decide what role you want to play in the life cycle of your device (concept to commercialization). Do you want to give it up for adoption (sell the IP early), or do you want to try to raise it to adulthood (obtain full regulatory approval to market), or do you want to participate something in between? That answer will help you decide if you need to conduct a clinical study on your device and if so, the type of study. Do not go into a clinical study lightly! You have an obligation to your subjects/patients to run the best possible study. They deserve to have their contribution count. Regulatory authorities demand accurate, truthful, and complete data. You must demand the same. After all, it is your reputation and that of your device on the line.

IMPORTANT POINTS

- A clinical feasibility study may be needed when only formal testing in humans can provide the evidence that the device offers the expected benefit (Proof of Concept); you will need data about the therapeutic effect *before* designing a larger pivotal study (or if you plan to sell the device, for the company planning to purchase your device before closing close the deal.

- Clinical trials are grouped into two basic varieties: significant risk (SR) studies and nonsignificant risk (NSR) studies; all SR studies require an IDE (investigational device exemption), while NSR studies only require IRB approval.
- If an IDE is necessary, an IDE Early Feasibility Study offers evidence of the amount of nonclinical testing required to support the IDE; the alternative is a traditional IDE which requires all nonclinical testing (bench and animal) to be completed prior to applying for an IDE.
- If *you* want to complete development of the device and obtain regulatory approval, meet with the appropriate authorities of the FDA to discuss your plans and obtain their advice. This presubmission or pre-IDE meeting helps determine what nonclinical testing the FDA will require, the endpoints and follow-up for the clinical study, and agreement of the design of your proposed clinical protocol, but you need to be fully prepared!
- Unless you are an experienced clinical trialist, do not try to manage a clinical trial yourself-hire a contract research organization (CRO) to manage the trial and one or more seasoned regulatory affairs professionals.
- Conducting a clinical trial entails many important concepts, some of which include: the number of patients needed, number of sites for the trial, a contract with each site, local IRB (Institutional Review Board) review and approval, a plan for interim monitoring visits, a clinical events committee, a Data Safety Monitoring Board, and possibly a core laboratory.
- You may need an experienced biostatistician with particular expertise in the analysis of complex study designs.
- All this is expensive! Be prepared for the costs included, not only for the trial (the most expensive part), but also if needed, a 510(k) application ($4690) and a premarket approval application (PMA) for a Class III device ($234,495).

Chapter 11

Innovating in a Rural Setting

Elizabeth Cole, James Hood, Dennis Matthews and Thomas Hobday
Tahoe Institute for Rural Health Research, Truckee, CA, United States

The region of North Lake Tahoe is a rural setting that serves as a very popular recreational area for many successful people from Silicon Valley and from the San Francisco and Sacramento areas of Northern California. But also within this rural community live some recognized scientists, physicians, and successful inventors, engineers, and entrepreneurs who have made their home in the community they have come to love; their goal has been to either retire here or interestingly, to continue to pursue their successful careers.

Within this rural region, Tahoe Forest Hospital, a designated Critical Access Hospital located in Truckee, California, is a rural health care facility. But this hospital also houses the Tahoe Institute for Rural Health Research (TIRHR), a subsidiary of the Tahoe Forest Hospital District. Within this rural setting, the mission of the TIRHR has been to develop working, functional, but novel practices/policies and innovative technologic products to deliver affordable but still high-quality health care not only to this community but also to similar remote, rural, underserved communities. A special focus of TIRHR has been to decrease the costs of health care delivery to rural clinics and hospitals in remote or home settings through innovative approaches.

TIRHR works with health care professionals with limited available resources or advanced medical technology primarily from remote or rural environments to develop projects to meet their needs in a cost-effective manner. The staff at TIRHR have a unique approach of collaborative interaction with medical professionals, experienced scientists, and business executives. TIRHR has a close working relationship with the NSF Center for Biophotonics, Science, and Technology at University of California at Davis (UCD), the UC Davis College of Engineering, and the UC Davis Health System. Via access to the scientists and laboratories in the UC Davis system, TIRHR uses their expertise to help to develop solutions needed in its efforts to increase access and decrease the costs of rural health care delivery. Equally important in the current world of innovation is the collaborative relationship that TIHR has developed with the Graduate School of Management at UC Davis in order to evaluate the marketability of some of the medical technologies it creates in this unique space.

Medical Innovation. https://doi.org/10.1016/B978-0-12-814926-3.00011-5

Although project ideas that are developed and originate in rural or even remote environments by TIRHR are designed to solve problems encountered by the different types of practitioners in remote areas, the innovative and often "disruptive technologies" under development provide for more accurate diagnostic and treatment outcomes for patients in all types of practices worldwide. These innovative systems and solutions must of necessity be portable, cost effective, and usable at home or in remote locations, and will often replace many of the more invasive diagnostic techniques used currently in urban medical settings.

The working approaches at TIRHR have many unique advantages. Interactions are managed by a Scientific Advisory Committee and its multidisciplinary Board of Directors. Medical professionals in the community identify needs and these lead to R&D projects. Once R&D projects are proposed, TIRHR conducts a thorough market analysis to determine commercial viability. This step has proved effective in avoiding interesting but improbable investments. Similarly, TIRHR also partners with selected, high quality, outside sources to meet its needs. As an example, the complex software needed for specific applications is subcontracted to experienced vendors in the appropriate field of medicine and innovation. A crucial component of the TIRHR organizational approach has been the mutually cooperative interaction and royalty-sharing policy with the infrastructure, available equipment, and talent at UC Davis. TIRHR provides management and exploration grants to the researchers at UC Davis to focus and accomplish the marketable goals of the projects rather than just to perform basic science research/discovery. UC Davis, in kind, readily provides access to their facilities, equipment, expertise in selected fields, and intellectual capital.

Innovation requires an environment of highly skilled people with ability and experience in solving problems creatively. TIRHR has taken specific advantage of the unique population of driven, successful, seasoned professionals from Silicon Valley, Sacramento, and the other nearby academic and business-oriented institutions in our community who have come to our area to "wind down" or to retire from highly successful careers where they have proven their skills at medical product and business innovation. TIRHR encouraged these unique individuals to participate in the development of innovative ideas to promote the introduction of better means of health care delivery that would also affect them directly. By starting on a volunteer basis, as the viability of the concept idea was established from both a scientific and business model of potential success, these bright, creative entrepreneurs were recruited to work on and contribute to the aspects of the project they were interested in, most often on a part time basis, for which they received a stipend plus partial ownership in any resulting product.

TIRHR is committed to focusing on and working to develop solutions to solving real problems identified by the medical professionals working in our community. Interestingly, although our development teams concentrate on innovative ideas that improve health care in a rural environment, all of the projects to date have also had applications to health care in general and even in urban settings. So many innovations in medical care that seem focused on one community also have applicability in other environments as well; innovative solutions to problems in small community or rural settings may also benefit the medical needs of patients in urban settings, because health care delivery is often deficient there as well. The environment created by TIRHR to encourage and stimulate innovation actively develops a structure which uses a multidisciplinary team of specialists in product design and development who work together with other groups at TIRHR to avoid creating a "silo" around each team. The interactions at regularly scheduled design reviews and full staff meetings allows cross fertilization of ideas and expertise across all the individual work groups.

TIRHR provides the "seed" funding typically for each project. Further relationships with outside funding resources are established later by each group, and importantly, it is this funding that takes the project through to its full development and commercialization. Although TIRHR serves as the start-up engine, TIRHR does not and cannot take the product through this latter process of commercialization. The skills and expertise of TIRHR are in developing the product from its concept verification to the prototype stage, but thereafter, TIRHR must partner with selected medical device companies to commercialize and then market the product.

Because of its business model, the need to "protect" the proprietary information associated with each project developed by TIRHR can be a difficult tradeoff while still maintaining the ultimate goal, that of developing the "science" behind each project and producing a viable product. One of the guiding principles of TIRHR is that qualified peer review is crucial to guarantee that the products or programs being developed are founded on solid scientific principles and will receive acceptance in the mainstream medical community. All projects are subjected to peer review along the various stages of development by such groups as the National Science Foundation, peer-reviewed publications, notable experts in the field, etc. We are careful to file patents prior to publication in an attempt to protect the intellectual property within each project. This process helps maintain an attractive environment of innovation for both our entrepreneurial partners as well as our scientific collaborators by allowing these development personnel to both maintain and enhance recognition within their business and/or scientific community.

Because of this business-oriented, product-driven, scientific approach, TIRHR has been able to capture the respect and interest of scientific experts within the National Science Foundation as a unique and extremely successful model for collaboration between industry and academia. The TIRHR Institute and its affiliated entity Cardiac Motion LLC, have sought and received three, peer-reviewed grants from the National Science Foundation to further develop several of the attractive projects at TIRHR.

In its short history of only 8 years, four US patent applications in various stages have been filed by TIRHR with the US Patent Office; an additional 14 international patents are also in process. As would be expected with a successful start-up engine like TIRHR, collaboration of our experienced staff with legal counsel was required to develop these patents.

To date, TIRHR's program has resulted in the development of four innovative projects with potential global impact.

PROTOCOLS AND ALGORITHMS FOR CRITICAL EVENTS

A practicing anesthesiologist in our community, Dr. Larry Silver, identified the need for a set of protocols and algorithms for the onset of critical events which led to the product we refer to as PACE (protocols and algorithms for critical events). The availability of a readily accessible set of critical care algorithms to serve as a cognitive aid to the clinician when confronted with a critical event occurring in an emergency situation could improve outcomes. The PACE application offers 30 separate protocols and algorithms to point-of-care environments developed for use by clinicians *during* the actual critical care event. PACE has proved to be invaluable during emergent situations in emergency rooms, ICUs, urgent care facilities, ambulatory surgery centers, and physician offices with independent, free-standing surgery suites; many of the physicians in these facilities are less well versed with handling these types of serious emergencies. Our PACE product helps to direct the physician to work through the appropriate decision tree algorithm for that specific event and after the event transmits a record of the actions taken to the patient's electronic medical record.

Currently, TIRHR is actively pursuing opportunities to commercialize this product. The system is being evaluated at Tahoe Forest Hospital, the UC Davis Medical Center, and Dignity Health in Northern California. TIRHR has filed for US patents on this system.

mTBI DIAGNOSIS SYSTEM

The mild traumatic brain injury (mTBI) research project had its origin after a conference of neurologists and scientists from Tahoe Forest Hospital and UC Davis, hosted by TIRHR, in Truckee, California in 2010. At that conference, the Senior Director of the Former Player Services of the NFL Players Association, Nolan Harrison, addressed the topic of concussions and challenged the conference participants to develop an objective method to diagnose a concussion and to determine when the brain has healed satisfactorily; this type of diagnostic device or test would markedly assist in the decision-making process related to either returning a player to the field or to prevent further potential injury to a brain not yet fully recovered. Mr. Harrison himself suffered from a concussion as a player in the National Football League as well as from the practice of returning players to the game without conducting an adequate neurologic exam. He argued that the systems available at that time to assist in the decision to return the player to the field were far too subjective. Players were often allowed to return to play before the brain was healed, which increases the risk of cumulative, long-term injuries.

TIRHR is working to develop and validate a portable, rapid, and inexpensive sideline mTBI detection system to provide objective evidence for the decision-making process for both removal from play and return to play. Personnel with minimal training, such as coaches, athletic trainers, and medical technicians, are the anticipated users of this system. In addition, the TIRHR Institute has partnered closely with the local school district, Feather River College, and the University of Nevada Reno on this project. TIRHR also sponsored a program in local schools to pay limited first dollar uncovered medical costs for student athletes >14 years of age to assure proper care after a concussion.

The eventual goal is to develop an objective test of <10 min that functions with 95% accuracy. Currently, albeit involving a limited number of patients, we have achieved an accuracy rate of 85%. We are confident that we will soon complete evaluation on an adequate number of patients through a proper period of time following the initial head trauma. Currently, we are testing patients after recovery to validate the ability of our system to determine when it is safe for the individual to the return to play. This project has been generously funded by the Gene Upshaw family through the annual Gene Upshaw Memorial Golf Tournament in Truckee.

THE PORTABLE BLOOD COUNT MONITOR

This innovative device provides a complete blood count (CBC), including the red blood cell count, platelet count, and white blood cell count (including the differential), the hematocrit, and the hemoglobin. The portable blood count monitor (PBCM) can also measure red and white blood cell counts in other body fluids automatically rather than requiring manual counting in a hemocytometer. Dr. Larry Heifetz, a practicing Oncologist in our area, identified the need for such a device that would require only a 3 μL blood sample which could be used by a patient, a physician's assistant, or a medical worker in a remote location, etc. The results can be transmitted directly to the treating physician within a few minutes. This device will allow one of the most fundamental but important medical tests to be obtained conveniently and immediately and reported to the point of care whether that will be in the physician's office, at home, or at a remote location anywhere there is phone or internet access. The device requires only a "finger stick" of 3 μL of blood.

Because the be PBCM will be totally automated, the health care professional or patient only needs to perform a finger prick, touch the slide to the blood, and insert the slide into the PBCM. The test results are available on site or transmitted to a separate location within 10 min. We have compared the measurements made by the PBCM to Clinical Laboratory Devices, and the results have been reported in several technical journals [1,2]. Similarly, measurements of red and white blood cell counts in a limited number of samples of peritoneal fluid, synovial (joint) fluid, and cerebrospinal fluid have demonstrated excellent correlation with hand-counted clinical results.

This project has been funded partially by two grants from the National Science Foundation, and we have filed for both US and international patents.

CARDIASCAN TM AMBULATORY VITAL SIGNS MONITOR

This device was designed to screen for dangerous cardiac arrhythmias in addition to simultaneously measuring the ECG, heart wall motion, and the rate, pattern, and tidal volume of breathing. This device allows surveillance of ambulatory patients over long periods for conditions such as paroxysmal atrial flutter and atrial and ventricular-based tachycardias. Future enhancements will measure changes in left/right ventricular volume during the cardiac cycle to monitor patients with heart failure and compromised cardiopulmonary function. Moreover, this technology is safe, easily and comfortably worn, nonobtrusive, noninvasive, and completely portable. The outputs from this device are uploaded to the cloud, thereby allowing a physician or skilled medical technician to monitor these records rapidly from patients in a remote location. Because this device measures both the physical motion of the heart and chest in sync with an ECG, this device differs from the current ambulatory monitoring devices like the Holter monitor.

As might be expected in our rural environment, the need for such a device arose from discussions with local cardiologists and ER doctors who requested a field-portable device to monitor patients with heart failure to diagnose possible arrhythmias as well as to follow cardiac output in order to allow optimization of medications.

At present, we have completed the proof-of-concept phase and will be developing a fully integrated, portable, prototype for clinical evaluation.

This project has also been funded in part by grants from the National Science Foundation and private investors, and we have filed for the appropriate US and international patents.

In summary, the TIRHR has shown that with appropriate planning, organization, and personnel, innovation in the field of medical devices can occur in an appropriate rural setting. Success requires partnerships with academia and industry, as well as funding from outside industry to take such "innovations" from concept to commercialization. An ongoing challenge involves getting "face time" with potential investors/partners from a rural location rather than when one is developing new products in proximity to the more usual locations of venture capital firms. We are finding with the appropriate personnel, even this hurdle can be overcome.

REFERENCES

[1] Smith ZJ, Gao T, Chu K, Lane SM, Matthews DL, Dwyre DM, Hood J, Tatsukawa K, Heifetz L, Wachsmann-Hogiu S. Single-step preparation and image-based counting of minute volumes of human blood. Lab Chip 2014;14:3029–36. https://doi.org/10.1039/c4lc00567h.

[2] Gao T, Smith ZJ, Lin T-Y, Holt DC, Lane SM, Matthews DL, Dwyre DM, Hood J, Wachsmann-Hogiu S. Smart and fast blood counting of trace volumes of body fluids from various mammalian species using a compact, custom-built microscope cytometer. Anal Chem 2015;87:11854–62. https://doi.org/10.1021/acs.analchem.5b03384.

Chapter 12

How Good Ideas Die: Understanding Common Pitfalls of Medtech Innovation

Katherine S. Blevins*, Dan E. Azagury*, James K. Wall*, Venita Chandra*, Elisabeth K. Wynne† and Thomas M. Krummel*

*Stanford University, Stanford, CA, United States, †Washington University School of Medicine, St. Louis, MO, United States

INTRODUCTION

As with all things, the probability of success in medical technology innovation lives somewhere between zero and one. Nothing in life is certain (save death and taxes); the innovator's goal in moving "ideas scribbled on a napkin" to safe, reliable, adoptable, and economically viable products for patient care is to sequentially "de-risk" a new technology, ever inching the probability of success closer to one. The trophy case of successes is readily apparent and widely touted. The graveyard of failures and their "postmortems" are featured less prominently. Hidden in the stories of both success and failure are the subtle confluences of technologies, time, investment, circumstance, and luck that lead any endeavor to its final resting place. Over the past 16 years, considerable experience has been generated at Stanford Byers Center for Biodesign on both outcomes. In this review, the authors have tried to extract key learning lessons.

ENHANCING THE PROBABILITY OF SUCCESS

Before considering how "good ideas die" it is worth quickly considering the other side of the coin—how good ideas survive, thrive, and flourish. We hope this point/counterpoint will be doubly illustrative.

We believe that there are several fundamental principles essential to increasing the probability of success. In the Silicon Valley/Stanford University ecosystem, there are a number of basic attributes that have consistently contributed to successful innovation over more than 80 years. These same themes are

Medical Innovation. https://doi.org/10.1016/B978-0-12-814926-3.00012-7

remarkably consistent among the top five universities with the greatest success in tech transfer [1]. Three key themes emerge in all.

1. A ***culture*** of both of collaboration and innovation.
2. A systematic ***commitment to technology transfer.***
3. An adjacent ***ecosystem*** of real world experts in implementation.

The ***culture*** of innovation in successful centers is created neither by accident nor by "fiat." Instead, enlightened leadership, prescient perception of opportunity, sustained experimentation with rewards for collaboration, and acceptance of thoughtful failures, all help to build a culture of people and ethos. Eventually, over decades, culture "permeates the air and the water."

Unwavering and sustained support for innovation must be pervasive throughout the institutional structure from the bench to the bedside; academic and administrative leaders, clinicians, patients, and industry partners are all essential to success. If any of these stakeholders are absent or uninvested, then tension, frustration, and failure ensues. The recent appearance—and acceptance—of a "clinician innovator" track in academic medicine may further enable and legitimize academic medical institutions in the translation of their discoveries to eventual products "in a box" to improve patient care.

Along with a culture of support, there must be a ***commitment to technology transfer***, moving from a discovery or an invention to a commercial reality. The boom in tech transfer emanating from universities was first fueled by the passage of the Bayh-Dole and Small Business Patent Procedures Act (P.I. 96-517) in 1980. Critically, this act changed the ownership paradigm fundamentally—ownership of intellectual property (IP) developed with federal funding was specifically transferred *from the government grantor to the university grantee*. Hailed by *The Economist* as "the most inspired piece of legislation of the 20th century" [2], this simple act has unlocked the symbiosis of faculty research teams, university tech transfer, and industry partners to achieve a shared goal of improving medical care. Indeed, investigators and universities have seen their work translated into utility in the public sphere with potential academic and societal benefit and perhaps some legitimate financial benefit. Industry is exposed to novel discoveries and inventions and brings critical know how of engineering, regulation, manufacturing, commercialization, and the subsequent partnership to the team to develop successful businesses around novel technologies.

The timely example of such a profound impact and its perceived value is the current debate for ownership of CRISPR-Cas9 technology and its application to the life sciences. By way of background concerning IP, the United States Patent and Trademark Office (USPTO), operated under a "first to invent" system until March 2013 with enactment of the America Invents Act [3]. This transition to a "first to file" system changed the fundamental aspects of how ownership and protection of IP is granted.

In the CRISPR-Cas9 case, University of California Berkeley and the Broad Institute of MIT and Harvard fought a high-profile court hearing over ownership of the patents for the gene-editing technology. Researchers at UC Berkeley had filed patent applications in 2012 disclosing how to use CRISPR for bacteria, but there were few details on how to use this in eukaryotic cells. Later in 2012, researchers at The Broad Institute of MIT and Harvard had filed patent applications on how to use the system in eukaryotic cells. In February 2017, a judge ruled that the patent applications were different enough that they did not preclude each other, making both valid [4–6].

How this will affect commercialization of the CRISPR editing tool remains to be seen. What is clear, however, is the enormous impact of technology transfer, universities recognition of the value of fundamental discovery and invention, and the perceived value of subsequent application to human health and disease.

The formation and continuation of the relationships between stakeholders in innovation is facilitated enormously by a well-established technology transfer office. Universities wishing to succeed in this arena must build and manage high functioning offices of technology transfer. Remember that such IP is developed within university classrooms and laboratories with the aid of taxpayer dollars, but with an attendant and legitimate societal expectation that societal benefits ensue. Using Stanford University as an example, the Office of Technology and Licensing (OTL) has received >$1 billion in income (over a 40-year period) from licensing inventions developed at Stanford University [7]. Direct revenue streams (licenses, royalties, and equity) as well as subsequent/attendant philanthropy and prestige associated with a successful invention (think the Page Rank algorithm at the heart of Google) are ascribed to the University. Without this contribution, the world would find information searches far more cumbersome! This reality would not be possible if technology licensing was difficult, inefficient, or muddled in the office of technology transfer.

Beyond ***culture*** and ***commitment to a robust mechanism of technology transfer***, there is a distinct advantage to working in an ***ecosystem*** that is supportive of innovation. Access to people, facilities, and financial capital are all essential, and will be elaborated on further. John Hennessy, the 10th president of Stanford University, was known for his "commitment to dissolving walls between disciplines, people, and ideas" [8]. The interdisciplinary collaboration not only applied to teams and projects within Stanford; he often described the "porous walls of the University" extending to the supportive, collaborative business ecosystem in Silicon Valley [9]. This does not mean that innovation cannot be successful in a region without prior successes; rather the probability of success is simply less in an area without abundant expertise.

WHAT DOESN'T WORK—HOW GOOD IDEAS DIE

We will now review lessons learned from failures. Although there is no specific "Secret Sauce," there are several common practices that reliably impede the probability of success. We will focus on five key points of failure.

- The lack of a deliberate process.
- Assuming invention = innovation.
- An incomplete or flawed team.
- Inadequate resources.
- Lack of cultural support.

The Lack of a Deliberate Process

There is a prevalent misconception that great inventions occur like a lightning bolt and with clarity and completeness. Nothing could be further from the truth. When an amazing innovation appears, it is easy to believe in a magic moment of insight. Unlike Archimedes, fresh from the bath running through the streets of Athens shouting "eureka!," seldom is there a single moment of epiphany. Berkun [10] explains: "The best way to think about epiphany is to imagine working on a jigsaw puzzle. When you put the last piece into place, is there anything special about that last piece or what you were wearing when you put it in? The only reason the last piece is significant is because of the other pieces you've already put into place. If you jumbled up the pieces a second time, any one of them could turn out to be the last magical piece. Epiphany works much the same way: it's the work before and after."

Gordon Gould, a primary inventor of the laser, had this to say about his own epiphany: "In the middle of one Saturday night, the whole thing suddenly popped into my head, but that flash of insight required 20 years of work I'd done in physics and optics to put all the bricks of that invention in there."

It is fair to say that no grand invention anywhere in history has escaped long, tedious hours required to take an insight and create it into something of utility for the world. It is the hard work, however unglamorous that may be, that matters. Epiphany is largely irrelevant, and a byproduct of the former. And the former all emanates from the unsolved problem, what we like to call "The Need."

"A well-characterized need is the DNA of a great invention" [11]. This mantra of the Stanford Biodesign process captures the importance of characterizing an identified need thoroughly. The "would be" innovator must observe, investigate, validate, and critique every aspect of the need. This characterization is nonlinear, iterative, and must be reexamined frequently as the process evolves. The need will have many facets, many stakeholders, and many entrenched parties, and understanding each of these is essential in the decision to pursue a given need or to move on to another. Time, energy, and resources are too scarce to squander.

The deep dive of characterization includes input and observations in unguarded moments from physicians, patients, nurses, technicians, and other staff. It involves understanding the anatomy, physiology, pathophysiology, and etiology around a disease state. The incidence and prevalence of a disease state, as well as the market around care episodes, must be understood.

Subpopulations that may allow for ease of entry into the market should be identified. The history of treatment options, alternatives, patient trajectories through the health care system, and previous innovation around the need must be charted. The regulatory pathways involved must be identified and plotted, including any clinical trials needed to achieve regulatory clearance or approval. The characterization is a dynamic process, and any good innovator has a "finger on the pulse" on all aspects of their need.

An increasingly important element of characterization is *value* [12]. In the past, any improvement was "de facto" economically viable: BUT NOT ANY LONGER!! Value in health care innovation is moving relentlessly toward proof that an innovation can deliver superior or equivalent outcomes with lower direct costs [13]. As the cost of health care comes under ever increasing scrutiny, the burden of proving that new innovations are of value shifts to the innovator/developer. This concept must be considered "a priori" [14] to any solution. Traditional avenues of payment through the generation of CPT codes, billing, and reimbursement are becoming more challenging to initiate and adjust. Coverage decisions are fluid as insurance marketplaces react to an uncertain future. Innovators must be creative in planning models of payment. Understanding the requirements, history, and current infrastructure around payment will help an innovator build an implementation plan which is fundable.

Inventing before completing need characterization is seductive but is a common mistake. This does not mean one should ignore insights and ideas during the process of needs characterization; rather, these ideas should be recorded and set aside for later analysis. Without the clearly defined specification of "must haves" and "desirables," an inventor can be lured by an elegant or clever solution that does not meet the criteria defined by the need. Linus Pauling said, "The best way to get a good idea is to get a lot of ideas" [15]. This approach is especially true in medical technology innovation. By establishing the need criteria, an interdisciplinary team has a methodology to objectively sort through many new concepts/ideas. Clearly describing these aspects of an identified need will lay the foundation on which to build idea generation, iteration, and invention.

Assuming Invention = Innovation

There is a continuum along which a new idea travels to become a commercialized product. In the book *They Made America* [16], Evans and colleagues observed "A scientist seeks understanding, an inventor a solution, and an innovator seeks a universal application by whatever means." Starting with discovery, a new bit of knowledge becomes a known. Such a discovery is often made in (but by no means always) a university, often supported by government or philanthropic grants. Such discovery occurs through an unpredictable and often serendipitous process. The application of these discoveries to patient care must be calculated given the consequences of error. Discovery thus becomes the fodder for the next step: invention.

Invention is an idea that builds frequently on preexisting concepts. Thomas Fogarty did not invent the balloon, and he did not invent the catheter, but he did invent the balloon catheter with initial application to arterial thrombectomy. An invention may not always be fully realized because of limitations of engineering or technology available at the time. Involved with invention is concept generation and validation through brainstorming, prototyping, testing, and iteration. The scoping of different embodiments of an invention allows for optimization of each element. This process, along with characterization through research in IP, regulatory, and reimbursement, can help the inventor choose a best design. None of this is sufficient for successful commercialization.

Innovation is the idea put into practice which proves to be useful and indeed valuable in effecting change. This is distinctly different from invention. It is the work, capital, and pursuit that push an invention to a realized, useful product. A brilliant idea without execution remains an idea ... indeed a hallucination [17].

The final step toward commercialization often requires an entrepreneur, that is, someone willing to take on the risk and work of a new enterprise and assume fundamental responsibility for its success. The execution of a theoretic plan is a challenge in of itself; the plan will need to adapt continuously, because crises abound. The skill, dedication, and work required to make this a success is often underestimated and is most often achieved by a team rather than a single individual.

An Incomplete or Flawed Team

The story of a lone inventor with an isolated moment of insight leading to a successful innovation is romanticized fiction. Flashes of insight result from years of work, research, study, and experimentation long before any "epiphany." With the exponential growth of human knowledge and the advances in technology and medicine, no single person can personally muster sufficient expertise—interdisciplinary teams have become essential. Such a push for interdisciplinary collaboration is seen throughout clinical, scientific, and corporate culture. So too, each innovator must not only contribute existing domain knowledge but also insight into the trends in their field. Some of the most successful companies have been started by at least two, and usually more, founders [18]. Successes in Silicon Valley often begin as a duo—Hewlett and Packard, Jobs and Wozniak, and Page and Brin to cite but a few. These duos often have complementary skills and are far more effective innovators because of it.

Any successful innovation/fledging business requires a resourceful leader who can enable teams to work together. As with the myth of the solo innovator, there is the myth of the perfect leader. "No leader is perfect. The best ones don't try to be—they concentrate on honing their strengths and find others who can make up for their limitations" [19]. This idea of the "incomplete leader" requires deep insight of said leader to recognize strengths and weaknesses and then build a team with complementary strengths.

Team dynamics and confidence will vary over time. Overconfidence can lead to critical mistakes, potential harm to the patient, and failure. Underconfidence can lead to project stall, resource depletion, paralysis, and again, failure. External coaching/mentoring can be formal (Board of Directors) and informal advisors, hence the value of the local ecosystem. A lack of effective oversight (occasionally referred to tongue in cheek as "adult supervision") distinctly lessens the probability of success. Absence of such oversight comes in many flavors—inadequate mentoring, "helicopter oversight," "Kool-Aid intoxication," lack of diverse mentors, or reliance on superficial mentors not totally invested in the success of their enterprise/mentees. The Theranos saga is a prime example of several such flaws [20].

Inadequate Resources

To bring any new technology from the bench to the bedside requires substantial resources. Innovators often define resources solely as investment capital. Throwing financial capital alone at a problem, however, only guarantees that capital will be spent!

The key capital currencies of innovation are *intellectual capital, risk capital,* and *financial capital* [21], each of which has its essential role in innovation. Lack of intellectual capital can leave a petri dish unexamined, an observation in a lab notebook undeveloped, unbuilt, and unimplemented, etc. Lack of risk capital can prevent the timid physician from developing as an entrepreneur. Perhaps most tangibly, lack of financial capital will ensure no idea will ever be resourced to pursuit.

The role of *intellectual capital* in commercialization is exemplified when reviewing the impact of Stanford University via discovery over a 75-year period [7]. Stanford has become one of the top tech transfer engines, demonstrated initially by Hewlett and Packard. Since then, there have been over 8000 inventions optioned generating $1.3 B in royalties. Technologies such as DNA cloning, FACS cell sorter, sound synthesis chips, DSL technology, and the PageRank algorithm were all discovered/invented at Stanford. There have been 39,900 companies formed, generating 5.4 million jobs and $2.7 trillion in annual revenue. Stanford alone would be the world's 10th largest economy. Stanford University has graduated the most entrepreneurs from their Undergraduate Education [22], even more impressive with just over 7000 undergraduate students annually. Simultaneously, more than 30,000 "not for profit" entities have also been birthed. This example is of a single institution! Imagine the outputs if this example could be replicated more broadly.

An equally critical capital is *Risk capital*—i.e., when innovators accept uncertainty both financially and in their career trajectory to pursue innovative avenues, often against the advice of experts. A historic example of risk capital can be seen in the story of Andy Grove. Andy grew up in German-occupied Hungary during World War II. His father was arrested and he and his mother took on false identities in order to escape. At age 20, Andy escaped Hungary, penniless and barely able to speak English, arriving in the United States in 1957. He earned his bachelor's degree in chemical engineering from City College of New York and then a PhD in chemical engineering from the University of California, Berkeley in 1963. He worked at Fairchild Semiconductor at the front of the microcomputer revolution. He then joined Robert Noyce and Gordon Moore at Intel as the third employee, ultimately serving as President, CEO, and chairman of the board for Intel. This exemplifies an absolute willingness to risk failure at every step of a life time [23].

In examining financial capital (and risk capital), one can look at Eugene Kleiner, another pioneer of Silicon Valley. Born in Austria, he fled Nazi persecution of Jews and arrived in New York City. He studied mechanical engineering and industrial engineering, then briefly taught engineering before joining Western Electric. He moved to California to help form what would become Shockley Semiconductor Laboratory in 1956, then left to found Fairchild Semiconductor in 1957. This then led to his own investment in Intel in 1968. He then joined Tom Perkins in 1972 to found Kleiner Perkins, later adding Brook Byers and Frank J. Caufield as named partners [24]. Kleiner, Perkins, Caufield, & Byers (KPCB), the Silicon Valley venture capital firm, has been an early investor in more than 300 information technology and biotech firms, including Amazon.com, AOL, Google, Genentech, and Twitter. KPCB, a mecca on the pilgrimage down Sand Hill Road for funding, still features "Kleiner's Laws" on their website, with such advice as "Invest in people, not just products" [25].

At different stages of development, demands for capital fluctuate. In need characterization and invention, intellectual capital is the most needed. Through the regulatory approval process, the demands for financial and intellectual capital increase markedly. Throughout the course of development, founders, inventors, and investors all invest risk capital. Daily developments in science/technology, newly entering competitors, the funding landscape, regulatory environment, and various avenues of reimbursement can cause an enormous shift in focus and capital needs for any innovation team.

Lack of Cultural Support

Innovation is often ignored as a "coin of the realm" in many universities, usually arising from the common myth that commercialization by an academic clinical innovator means less academic productivity. To the contrary, there is an abundance of evidence supporting the correlation between academic productivity and innovation. Ahmadpoor recently documented that most cited research articles link forward to a future patent (80%). Similarly, most patents (61%) link backwards to a prior research article [26]. This observation is consistent with the theory that there are powerful connections between patents and scientific inquiry. There are multiple other correlations between innovation-commercialization and impact in scientific publications [27–30]. The evidence supports that the academic clinical innovator should be a valued faculty member in academic medicine.

The clinical innovator path is indeed different than a traditional academic career path in "bench science." Many "nontraditional" pathways are appearing that combine clinical practice and different aspects of inquiry. Some examples include translational research, prevention research, outcomes research, education research, and policy. Academic medical institutions can/should be well positioned to serve as incubators of innovation [31,32]. Traditional metrics of success, such as grant funding and publications, may need to be balanced with nontraditional metrics (advisory roles, patents, software licenses, etc.) for advancement and promotion if Clinician-Innovators are to survive, thrive, and flourish.

CONCLUSIONS

As outlined above, there are several key elements that push an innovation toward success: a culture that supports innovation, a mechanism for commercialization through a supportive technology transfer process, and a regional milieu/ecosystem that gives a distinct advantage. Conversely, we reviewed multiple ways to kill a good idea such as:

1. Lack of process in innovation.
2. Assuming that once an invention is made the work is complete.
3. Having an incomplete team.
4. Lack of a variety of resources.
5. Limited cultural support around innovation.

By no means do these lessons learned create an absolute formula for success. We do believe that an awareness of these criteria can increase the chances of success for future academic innovators. Success in MedTech Innovation is a fickle, temperamental, moving target. A successful innovator must do everything he or she can to increase—or at least not decrease—that probability.

IMPORTANT POINTS

- Success in medical technology innovation is challenging, however, there are key principles that can increase or decrease the probability of success.
- Common principles that increase the probability of success include a culture of collaboration and innovation, a commitment to technology transfer, and an adjacent ecosystem.
- Conversely, the principles that lower the probability of success include the lack of a deliberate process, the assumption that invention = innovation, having an incomplete or flawed team, having inadequate resources, and the lack of cultural support.

- These key principles alone are insufficient to ensure success; however, the awareness of these common themes can help direct the innovator toward a successful endeavor.

REFERENCES

[1] Schramm C. Five universities you can do business with. Inc.com; 2006. Available at, https://www.inc.com/magazine/20060201/views-opinion.html.

[2] Innovation's golden goose. The Economist, Technology Quarterly: Q4 2002.

[3] USPTO. America Invents Act: effective dates, https://www.uspto.gov/sites/default/files/aia_implementation/aia-effective-dates.pdf; 2011.

[4] Cohen J. How the battle lines over CRISPR were drawn. Science 2017. AAAS, Available at, http://www.sciencemag.org/news/2017/02/how-battle-lines-over-crispr-were-drawn.

[5] Contreras JL, Sherkow JS. CRISPR, surrogate licensing, and scientific discovery. Science 2017;355:698–700.

[6] Ledford H. Broad Institute wins bitter battle over CRISPR patents. Nat News 2017;542:401.

[7] Eesley C, Miller W. Impact: Stanford University's economic impact via innovation and entrepreneurship. Stanford, CA: Stanford University; 2012, https://doi.org/10.2139/ssrn.2227460.

[8] Stanford 125. Transformative leadership: John Hennessy. Stanford 125; 2016. Available at, http://125.stanford.edu/hennessy/.

[9] Personal Communication Between John Hennessy and Thomas Krummel.

[10] Berkun S. The myths of innovation. Sebastopol, CA: O'Reilly Media; 2007.

[11] Yock P. Needs-based innovation: the biodesign process. BMJ Innov 2015;1:3.

[12] Fuchs VR. New priorities for future biomedical innovations. N Engl J Med 2010;363:704–6.

[13] Porter ME. What is value in health care? N Engl J Med 2010;363:2477–81.

[14] Gottlieb S, Makower J. A role for entrepreneurs. Am J Prev Med 2013;44:S43–7.

[15] Harker D. Letter to Linus Pauling; 1961.

[16] Evans H, Buckland G, Lefer D. They made America: from the steam engine to the search engine: two centuries of innovators. New York, NY: Back Bay Books; 2009.

[17] Wynne EK, Krummel TM. Innovation within a university setting. Surgery 2016;160:1427–31.

[18] Kawasaki G. The art of the start: the time tested, battle-hardened guide for anyone starting anything. New York, NY: Portfolio; 2004.

[19] Ancona D, Malone TW, Orlikowski WJ, Senge PM. In praise of the incomplete leader. Harv Bus Rev 2007;85:92–100. 156.

[20] Paradis NA. The rise and fall of theranos. Sci Am 2016. Available at, https://www.scientificamerican.com/article/the-rise-and-fall-of-theranos/.

[21] Krummel TM, et al. Intellectual property and royalty streams in academic departments: myths and realities. Surgery 2008;143:183–91.

[22] Stanford K, Hammond R, Sam J. PitchBook top 50 universities report: the top 50 universities producing VC-backed entrepreneurs. PitchBook; 2017. https://pitchbook.com/news/reports/2017-universities-report.

[23] Isaacson W. The innovators: how a group of hackers, geniuses, and geeks created the digital revolution. New York, NY: Simon & Schuster; 2014.

[24] Kleiner E. Obituary: Eugene Kleiner. The Economist December 6, 2003;.

[25] Eugene Kleiner — Kleiner Perkins Caufield Byers. Available at, http://www.kpcb.com/partner/eugene-kleiner. Accessed 23 August 2017.

[26] Ahmadpoor M, Jones BF. The dual frontier: patented inventions and prior scientific advance. Science 2017;357:583–7.

[27] Azoulay P, Ding W, Stuart T. The impact of academic patenting on the rate, quality and direction of (public) research output. J Ind Econ 2009;57:637–76.

[28] Breschi S, Lissoni F, Montobbio F. University patenting and scientific productivity: a quantitative study of Italian academic inventors. Eur Manag Rev 2008;5:91–109.

[29] Van Looy B, Callaert J, Debackere K. Publication and patent behavior of academic researchers: conflicting, reinforcing or merely co-existing? Research Policy 2006;35(4):596–608. Available at: https://ssrn.com/abstract=1502458.

[30] Thursby M, Thursby J, Gupta-Mukherjee S. Are there real effects of licensing on academic research? A life cycle view, National Bureau of Economic Research; 2005. https://doi.org/10.3386/w11497, http://www.nber.org/papers/w11497.

[31] Majmudar MD, Harrington RA, Brown NJ, Graham G, McConnell MV. Clinician innovator: a novel career path in academic medicine: a presidentially commissioned article from the American Heart Association. J Am Heart Assoc 2015;4:e001990.

[32] Yeo HL, Kaushal R, Kern LH. The adoption of surgical innovations at academic versus nonacademic health centers. Acad Med J Assoc Am Med Coll 2017. https://doi.org/10.1097/ACM.0000000000001932 (e-pub ahead of print).

Chapter 13

Managing Institutional Barriers to Entrepreneurship

Thomas R. Mackie* and Eric Leuthardt†
**University of Wisconsin, Madison, WI, United States, †Washington University, St. Louis, MO, United States*

Entrepreneurship by academicians is not new. San Jose and South San Francisco Bay was a center of California orange production at the time of founding of Hewlett-Packard in 1939 by Stanford graduates William Hewlett and David Packard [1]. Hewlett and Packard were encouraged to start their own companies by Dr. Fred Terman, a professor of electrical engineering who studied under Vannevar Bush and who personally invested in his students' spin-off companies, including Litton Industries and Varian. He was the prime mover of the Stanford Research Park that leased land to Stanford spin-offs as well as Kodak, General Electric, and Lockheed. From 1955 to 1965, Professor Terman was Provost of Stanford, guiding his university to become one of the most entrepreneurial in the world.

After World War II, Americans understood that science and technology had contributed greatly to the war effort, and university stature had helped establish its place as the leader of the free world. American universities rose quickly to be the leading institutions in the world due to a brain drain from a crippled Europe and generous federal funding of universities in the United States. Industry looked to universities to help their companies. Universities began patenting and started technology transfer offices to manage intellectual property (an interesting exception is the University of Wisconsin where faculty started an independent, not-for-profit technology transfer operation, Wisconsin Alumni Research Foundation, in 1925). Ready federal funding with grant money and a huge expansion in the student population in the 1950s and 1960s sharpened the focus of universities on the metrics of state support per student, tuition, and indirect proceeds from grants. In the past 50 years, state support for public universities has decreased in real dollars per student, while tuition has increased by an order of magnitude, and the proportion of federal share of national research spending per GDP [2] has decreased from 2/3 to 25%, while industry research spending per GDP has increased from 1/3 to 2/3. Industry and especially

Medical Innovation. https://doi.org/10.1016/B978-0-12-814926-3.00013-9

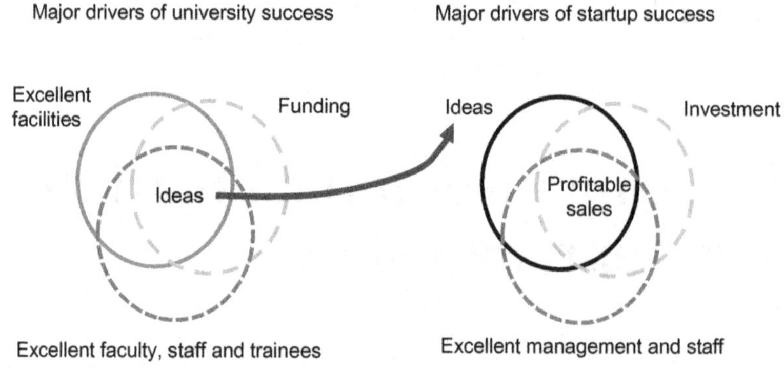

FIG. 13.1 The symbiosis of universities and university start-up companies.

start-ups now rely on the number one product of universities: ideas. As illustrated in Fig. 13.1, the overlap of ideas along with investment and capable management and staff are the main determinant of start-up success. Graduating students and other trainees, if lucky enough not to be saddled by student debt, are at a point in their lives with the greatest talent-to-caution ratio—ideal for a founding team. It is no wonder that universities are now increasingly associated with high-tech start-up communities.

There are no American universities today which forbid academic entrepreneurship, and indeed, the very best universities are encouraging it actively. All universities have well-defined policies of conflict of interest which manage the forays of faculty and staff into business as discussed in other chapters in this book. Large universities have a research park that encourages both start-ups and existing companies to lease land or buildings to get access to expertise at their university. All universities that obtain a large fraction of their support from federal grants have flourishing technology transfer offices that take invention disclosures from academicians, file patents, license technology, and encourage start-ups. Some offer opportunities of seed investment for university start-ups often funded by alumni. Nevertheless, the culture of entrepreneurship at most universities, while improving, can often be described as benign neglect in most of their schools and colleges; indeed, residual hostility is still present at some medical schools. While administration, faculty, and certainly students understand the opportunities at many institutions, administrative levels of dean, center directors, and department chairs often put entrepreneurship below community service in value. Students trained and indirect grant support remains the coin of the realm to university middle management. Medical schools that have partnered with their practice or insurance plans, high revenue businesses in their own right, deserve special discussion.

Language is the root of cultural understanding and so too with entrepreneurship. The phrase "conflict of interest" connotes unseemliness, if not outright corruption. Federal statutes mandate that "conflict of interest" committees manage academic entrepreneurship. A simple boost to entrepreneurial culture would be to have these committees renamed "potential conflict of interest" or better yet, "conflict management." The latter would be best, because there are few worthwhile human endeavors that are devoid of some degree of possible conflict, but the key is to manage them so that progress can be made and not hindered. Unfortunately, the "pharmaphobia narrative" [3], vigorously embraced especially at medical schools (see the Chapter 14 by Stossel) presupposes that academic-industrial relations are fraught and intertwined with danger, and prevailing academic policies haplessly attempt to remove all potential commercial conflicts [4]. This outlook and too many of these policies actually and functually serve to impede progress in medicine and engineering. A world denuded of potential conflict is the ultimate nanny state, wonderful for bureaucrats but awful for both burgeoning innovators and established visionaries.

Too often, "conflict of interest" is confused with "nexus of interest." A nexus of interest does not have to be in conflict. There can be a coincidence of interest without conflict. A good example is the way that many medical innovators and inventors are treated by their Institutional Review Board (IRB) who manages the clinical trials. The IRBs at most universities require a waiver for an inventor to be involved with a clinical trial involving his or her own technology. Usually the inventor is the world's expert on said technology; to have them uninvolved in the technical design of the trial is dangerous, because they typically know the most about the indication being tested and what problems can arise. If they are the expert, then they are often excluded from giving sufficient input into the trial design. The patients, the inventor, the company that licensed the invention, and the medical school all have a coincidence of interest to ensure that the trial is safe and that the intervention can be tested effectively. The potential conflict of interest only arises at the operational stage, that is, the selection of patients, the care of the patients, the analysis of the results, and their publication. The operational stage can usually, but not always, be supervised by someone who reports independently to the department chair or designated alternate without a potential conflict of interest, but the design of the trial must have the proper input by technical experts. Indeed, instead of a default prohibition of involvement of the inventor, perhaps there should be a default inclusion. In that scenario, the inventor would have to affirm that they do not want to be involved with the technical design and that, in their opinion, there is sufficient technical expertise to design the trial safely. This would rarely be necessary. The inventor naturally has strong incentives to protect the patient, logically wants to offer suggestions on design, and any objective trial supervisor should naturally be receptive to the ideas of the inventor, that is, the true expert. Again, this is because there is a coincidence of interest that drives everyone involved to want to conduct clinical trials as safely as possible. Ultimately, it is hoped that governments which have promoted suspicion toward the medical industry will become more comfortable with innovation and technology transfer from university to industry. Then, the inventor will be presumed innocent of corruption and be able to conduct all aspects of the clinical trial just as university research faculty are presumed not to have biased their research. Jurisdictions which make this leap of faith demonstrate a firm commitment to attracting the best medical innovators.

Another example of the difference between a conflict of interest and a nexus of interest involves continuing education with medical professionals. The patient wants the physician to have thorough training on any new technology to which the patient will be subjected and especially any technology capable of causing harm. The company selling the technology wants their product to be used properly, because this maximizes and predicts favorable clinical outcomes and a positive reception by the market. The hospital wants its staff to be as competent as possible, minimizing legal, economic, and brand risks. There is no conflict, indeed only a coincidence of interest by all to the maximum level of performance from the technology, team, and institution. Unfortunately, the typical policy calls for medical schools to prohibit their practitioners from attending Continuing Medical Education (CME) that has not been certified by the Accreditation Council on Continuing Medical Education (ACCME). CME credits are used by doctors and other medical professionals to certify that they are competent to practice medicine. Getting ACCME CME accreditation is costly and time-consuming for industry and is only used for larger symposia. Ironically, local training by industry representatives can be done if the hospital pays fair market value for the training service; applications trainers typically spend days to weeks working with hospital staff. In orthopedic surgery for example, company representatives are often key members of the team working in the surgery suite at the side of the surgeon. It remains a mystery why such intimate, long-term association with industry representatives is not a conflict of interest just because the hospital pays for the education service to the company employing the trainer. Industry does not decry this situation, because applications services are profitable. Ironically, interactions between industry and the deliverers of health care are said to be regulated in practice to improve patient health and decrease spending on health care [5]. Professional services and non-CME accredited presentations are unlikely to brainwash academicians and medical professionals, because these clinicians and academics are highly educated, tend to be naturally skeptical, and can readily separate the hyperbole of industry sales from valuable product information. The potential conflict of a drug or medical device company encouraging the use of their product where it is not effective is best dealt with by targeted payment exclusion by payers and not prohibition of information to providers. It is interesting to note that this may be one of the only educational experiences highly regulated by our universities. Outside speakers on campus can espouse outlandish theories freely and many allow hurtful pedagogies on religion, race, gender, and sexual orientation to undergraduates on the reason of exposing the students to a wide range of ideas. Why should we feel the need to protect doctors, typically with more than 10 years of university education, from medical industry representatives?

Conflict of commitment, here called "commitment management," is a topic related closely to, but different than, conflict management. The "day-a-week" allowance for "outside activities" by an academic entrepreneur is an unwritten policy at many universities, but there are marked differences between colleges and departments on the amount of effort allowed for these activities. The National Endowment for the Arts [6] reports that fine arts and music faculty at a typical university are entrepreneurs, and many arts business programs are taught in connection with art and music departments and business schools. There are three reasons why academic artists do not live in ivory towers. First, artistic faculty are generally poorly paid, and outside income is needed. Second, even if faculty salaries in the arts were to increase commensurately with the true worth of the artists, it is to the benefit of the university that the artists engage with society commercially. Finally, artists are judged at the time of their hire and by their tenure committees on their artistic influence on culture. The litmus test for cultural influence is if the artists are able to sell their work or attract an audience; in most cases, fine arts faculty were established entrepreneurs *before* they were professors. Entrepreneurialism in the fine arts is also treated differently. A painter allows their students to watch them paint with materials paid for by a grant from a foundation. Neither the university nor the foundation would restrict these artists from selling their work in their off-campus private studio. A professor of music performance would be expected to play for a professional symphony or go on tour. The story is similar in the humanities. History professors would not be expected to turn over the royalties of their textbook to the university, and the university would not question the motives of these professors if they spent their office hours drafting the tome. In these cases, the university would celebrate the success of these artists and humanitarians to their alumni. The commitment that the universities expect from their artistic and humanitarian faculty explicitly includes their outside engagements; indeed, the blurring of inside and outside activities are celebrated by their universities.

While the level of time commitment to outside activities is different across the university as a whole, the decisions embedded in the guidelines should be consistent. The medical school perhaps cannot afford to have a transplant surgeon spending a day a week on entrepreneurial or consulting activities. Likely, the transplant surgeon would also agree for two reasons. First, at a majority of institutions, the university medical foundation and the transplant surgeons would get paid much more for clinical service than for their other academic commitments. Secondly, any other activity would likely not be as beneficial to society as their surgical skill. But, for the majority of faculty in an engineering or medical school, their technology or discoveries that are transferred to industry will help many more people than they could diagnose or treat themselves. Many of these faculty believe that there is more to nonclinical, academic medicine than teaching more students, getting another grant, or writing another article, especially when they are working on an innovation that may affect many more patients than the teaching of medical students may impact. In light of many scientific and clinical breakthroughs that were conceived and validated preliminarily in the university, including such innovations as X-ray and magnetic resonance imaging, angioplasty and stenting, and vaccines and antibiotics, history suggests the aspirations of academic inventors are well founded.

When a department chair or dean approves the entrepreneurial activity with an agreed amount of time commitment, admittedly the university is taking some risk. It is a similar risk, though different in degree and type, which the university takes when they allow a professor to teach an unpopular concept or enable a virologist to house and study dangerous viruses. The university is not sanctioning the concept being taught nor saying that the probability of the virus escaping is zero. They are taking a measure of risk for a much greater possible reward; in the language of medicine, the outcomes are expected to greatly exceed the complications. For every connection that a university faculty member makes with industry or the local entrepreneurial ecosystem, their university is strengthened. Rather than viewing academic entrepreneurship as an entanglement, universities should see and acknowledge this form of outreach to society as an engagement.

By approving the time commitment for entrepreneurship, research partnering with industry, or consulting, the university accepts that activity as part of its mission. For that faculty member, entrepreneurialism becomes as legitimate as teaching, research, and clinical service. In fact, the so-called "outside" activity has been rendered an "inside" activity. With that comes certain necessary obligations of the university and of the entrepreneurial faculty member. Both take time for managing potential conflicts. Faculty members are made keenly aware of disclosure of their entrepreneurial efforts in oral and written communications. Their students and colleagues are also informed. Efforts to ensure that the results of research, especially by students and other trainees, are published without undue delay (examples of short term, acceptable delays are reviews for invention disclosure and giving short-term advance notice of publications to industrial sponsors). Entrepreneurial faculty members cannot influence the university unduly concerning possible contracts with their enterprise or companies with which they consult. Clinical inventors and their institution should receive no royalty payments related to products invented in the institution and being used on the patients they care for under experimental conditions. A few simple rules can go a long way toward managing conflicts, but it takes good will and dialogue between an open academic entrepreneur and a supportive university administration to ensure that unacceptable conflicts are rare and of small consequence when they occur.

There are several policies which stem from establishing a level of commitment for the academic faculty member to engage in entrepreneurial activities. Similar to research and teaching activities that use unmetered resources such as offices, labs, computers, Internet access, and local telephones, entrepreneurship should also be treated in the same fashion, reflecting the allowed, negotiated commitment for entrepreneurial endeavors. Allowing *de minimus* use of university facilities creates a culture and climate of entrepreneurship on campus which signals to faculty and staff that academic entrepreneurship can be a legitimate university activity. Enabling accounting procedures so that metered use of resources can be treated in the same way as for research, teaching, and clinical service activities will improve efficiency. For example, travel and accommodation for combined business and university activities may be fairly separated.

Academic entrepreneurship also extends to training and education. The longer a trainee remains at a university, the greater the focus of training is directed at becoming a professor. The pinnacle of training of PhD or medical students is a postdoctoral fellowship or a medical residency. The truth is that most postdoc PhDs or residents will never be professors. According to a UCSF study, about 20% of postdocs will work for industry if for no other reason than most research and development is paid for by industry and only 25% by the federal government [2,7]. Because there are more than 5000 hospitals in the United States and fewer than 200 medical schools, most residents will work for a private clinical provider and not an academic hospital. Universities must prepare trainees properly for life outside of a university and not just exploit them as cheap labor on soft and shrinking federal money in the guise that they are training them for the next generation of professors. The training in business and entrepreneurism should not be compulsory, but it should be solely at the discretion of the trainee without influence from their supervisors. Trainees with some business training are also ready conduits for high-tech university spin-offs. It is the authors' experience that the most successful university spin-offs have a postdoc as Chief Technology Officer who trained in the lab of the inventor of the technology. Short courses on entrepreneurship and business offered to postdocs, residents, graduate students, and even undergraduates will also attract the best students and trainees to the university. These contentions are echoed in several other chapters in this book by medical student entrepreneurs as well as entrepreneurs in their medical residency years.

The most important cultural change a university can adopt to encourage entrepreneurship is for university administrators, deans, center directors, and department chairs to keep metrics on entrepreneurship and industrial engagement. Administrators now keep metrics on the number of students taught, the average time to get a first degree, the number of publications and their citation indices, the grants awarded, and the direct and indirect funding provided. These administrators should also be accountable for metrics on the number of invention disclosures and licenses granted, the number of spin-off companies and the investment they raise, as well as the number of faculty who are consulting for industry. There are reasons to suspect that technology transfer offices of universities do not have accurate data on start-up and consulting activities. The technology transfer offices of many universities do attempt to keep track of academic entrepreneurship, but the numbers are often siloed out of view of administrators, and conflict management is performed by departments and colleges so they have the closest view of this activity. The Association of University Technology Managers (AUTM), an association of university technology transfer employees, estimated that US universities produced 1024 start-ups in 2015 [8]. In the same year, $39.2 billion was spent on federal research. The AUTM data would imply one start-up generated per $38 million spent on federal R&D. This is dramatically contradicted by an independent survey in 2011; the National Cancer Institute (NCI), US Patent Trademark Office, and direct surveys of scientists found that one in four oncologists who had NCI grants have started a company and that there was one start-up per $12 M spent on oncology R&D [9]. Recently, the Milken Institute published a ranked list of university technology transfer activities [10]. It scored 225 American universities on the basis of their formation of start-up companies as well as their success at patenting and licensing. Their information was compiled from university technology transfer offices, and it is highly likely that many start-ups and consulting businesses were missed, especially those not started by STEM (science, technology, engineering, medicine) faculty and by staff and students. If such reports were compiled annually with input compiled by departments and colleges and not just technology transfer offices, public universities would find an increase in entrepreneurial activities and industrial engagement as a justification for state funding. Because university administrators excel at applied game theory, very quickly entrepreneurism and industrial engagement will be one more measure with which success can be celebrated. It is highly likely that the reported cases of entrepreneurship would increase dramatically because of more accurate reporting but also because of a change in culture that collecting and analyzing the data would engender.

A university that truly wants to be considered entrepreneurial should declare itself so. Aalborg University in Denmark has declared that entrepreneurship is integral to its core strategy [11]:

> *Hence, entrepreneurship is an integral part of [Aalborg's] core strategy and elemental to its innovation approach as a knowledge-generating and culture-bearing institution that contributes to technological, economic, social, and cultural innovation.*

It is important to remember that the commitment embraced by the university to activities in economic development would be but one measure of the university's engagement in society. At most, entrepreneurial universities will have only a minority of faculty and staff who will engage in entrepreneurship. But, those who found or assist start-ups should receive full credit for their entrepreneurial activities. Committees on promotion and tenure should acknowledge and apply the same weighting to productive invention as teaching and research. Most researchers will still be driven solely by curiosity, not application, but more would be aware of the possible benefit of their research to society. Few teachers will give classes on business principles in health care, although that scenario is changing if for no other reason that academic medicine is becoming seen through business eyes. Most clinicians will spend the vast majority of their time treating patients; entrepreneurialism will not fundamentally change the character of the university but will make it more relevant to society.

In summary, here are both important points and guiding principles and recommendations if universities want to embrace entrepreneurialism:

1. A potential conflict of interest does not mean a conflict actually exists. The university should look for "coincidences of interest" in the nexus of legitimate activities.
2. Engagement between industry representatives and academics should be encouraged actively.
3. An agreement with the university that the investigator may spend time on entrepreneurial activities, as described in the investigator's management plan, carries with it a responsibility of the university to embrace the activity. In turn, the investigator must also be responsive to the plan laid out by the campus Conflict Management Committee.
4. Trainees should have the opportunity, at their sole discretion, to take courses on business and entrepreneurism to prepare them for a life beyond the university.
5. *De minimus* use of unmetered university resources and the treatment of scarce metered resources should be treated like other legitimate university activities.
6. Being the most capable technical experts, academic inventors or entrepreneurs who develop a potential medical technology to be used in a clinical trial should be encouraged to participate fully in the design of that trial even though prohibited from leading the operational and reporting aspects of the trial.
7. Deans, center directors, and department chairs should report to the technology transfer office and the university administration all relevant metrics of entrepreneurship, and these findings should be made public.
8. Faculty entrepreneurship outside of the STEM fields, including the humanities and fine arts, indicates that entrepreneurship is widespread in a university and can provide another uniting framework for the university mission.
9. Universities should declare themselves as Entrepreneurial and fully embrace the importance of economic engagement with society as a mission alongside teaching, research, clinical care, and community service to further improve the economy and health care in the United States.

FINANCIAL DISCLOSURES

TRM has a financial interest in, is on the board of, or within the last five years has consulted or advised for multiple companies including Accuray, Asto CT, BioIonix, Cellectar, HealthMyne, Image Mover, Integrated Vital Medical

Dynamics, Leo Cancer Care, Oncora, OnLume, Redox, Shine Medical Technologies, and SmartUQ. ECL has a financial interest in, is on the board, or has consulted for multiple companies including Neurolutions, Osteovantage, Face to Face Biometrics, Pear Therapeutics, General Sensing, Inner Cosmos, Medtronic, Ascension Health Ventures, Greysy Ventures, Alcyone, and E15.

REFERENCES

[1] Gilmore CS. Fred Terman at Stanford: building a discipline, a University and Silicon Valley. Redwood City, CA: Stanford University Press; 2004.

[2] Press WH. What's so special about science: (and how much should we spend on it?). Science 2013;342(15):817–22.

[3] Stossel T. Pharmaphobia. Lanham, MD: Rowman & Littlefield Publishers; 2015.

[4] Lo B, Field MJ. (EDs.), Institute of medicine report on conflict of interest in medical research, education, and practice. Washington. DC: The National Academies Press; 2009.

[5] C. Kimmelstiel, Restrictions on interactions between doctors and industry could ultimately hurt patients, *J Vasc Surg* 54(18S):12-14.

[6] Sidford H, Frasz A. Creativity connects: trends and conditions affecting U.S. artists. In: National Endowment for the Arts. Washington, DC. 2016.

[7] https://postdocs.ucsf.edu/news/career-outcomes.

[8] Highlights of AUTM's U.S. licensing activity survey FY2015. Association of University Technology Managers, Washington, DC. 2016.

[9] Aldridgea TT, Audretscha D. The Bayh-Dole Act and scientist entrepreneurship. Res Policy 2011;40:1058–67.

[10] DeVol R, Lee J, Ratnatunga M. Concept to commercialization: the best universities for technology transfer. Santa Monica, CA: Milken Institute 2017.

[11] https://heinnovate.eu/en/resource/pathways-entrepreneurs-aalborg-university.

Chapter 14

Concerns About the Current Pharmaphobia in the World of Innovation: Its Consequences and Risks

Thomas P. Stossel[*,†]
[*]*BioAegis Therapeutics, Inc., Belmont, MA, United States,* [†]*Options for Children in Zambia, Foxboro, MA, United States*

Currently in the world of research and innovation, there exists an unjustified mindset hostile to private industry—"innovation pharmaphobia"—that has and continues to hinder the development of truly innovative advances in health care. This phenomenon is interesting but unfortunately is sad, because most translational applications of true discovery occur in the pharmaceutical and medical device industries [1,2]. Because of this reality, some form of partnership with industry is not only smart and scientifically necessary because of the ability to share in experience and innovation, but also to address the fact that federal research funding is not keeping up with innovation opportunity.

When examined from the aspect of academic productivity and practical innovation, partnering with industry allows the sharing of ideas, more funding for discovery and exploratory research, and of course the sharing of the ultimate market value of any new ideas that are developed and marketed successfully.

Society wants innovation and, for these reasons, it invests in many aspects of innovation with the goal of developing new and improved health care devices and medications to improve longevity and quality of life. While the house of academia touts the idea that discovery and innovation originates in the Ivory Towers of its world of secondary education and its health care institutions [3], the truth is that since the middle of the last century, most clinically useful innovations have had their birth in the world of industry where many of our really smart and well-trained PhDs and physicians "escape" to pursue their dreams and innovative thoughts [1,2]. It is only via the sharing of knowledge and experience that allows the unfettered development of new therapies in the real world as discussed in this book.

Medical Innovation. https://doi.org/10.1016/B978-0-12-814926-3.00014-0

Indeed, the direction of academic discovery into the market often requires interaction between basic science investigators/inventors and the expertise and practical experience of industry to allow translation of innovation into the public realm; this translation justifies public investment in research in the academic world. One can argue that it might even be considered unethical NOT to facilitate practical applications of research into the public sector.

In our current political environment characterized by a prevalent animus against big business, a phenomenon I have called "the pharmaphobia narrative" has arisen. This pharmaphobia has led to a ubiquitous morass of burdensome and stifling regulations concerning "conflict of interest" based on the assumption that any sponsorship with industry will lead directly to the misrepresentation of many of the results of research to fill the financial coffers of industry; in some cases, these accusations have even been directed back to the original innovators/inventors in academia.

According to the pharmaphobia narrative, partnerships with industry incentivize inappropriate use of medical devices or prescribing of new and costly medications of little additive value to current therapy just to reimburse the sponsors financially at the expense of the public. To deal with these most often unfounded allegations, both academic institutions and government have erected bureaucratic agencies to "manage" or even prevent potentially productive relationships between industry and the investigator/inventor/innovator when concerns are voiced about inappropriate interactions by naive or all too often uninformed administrators who often have an upfront bias.

While yes, some unethical interactions in the past between industry and investigators have occurred, these are by far the minority. Overall, however, I maintain that the plethora of governmental and academic regulations have and continue to stifle industrial partnerships with excellent, potentially successful, well-meaning research in academia.

In a book I have written entitled "Pharmaphobia: How the Conflict of Interest Myth Undermines American Medical Innovation" [4], I address the empiric basis of this theory with the conclusion that this pharmaphobia is both paranoid and unjustified. The often useless, time-consuming, and stifling roadblocks and regulations serve to force aspiring innovators who very well may be basically naive about business interactions to learn and comply with all these rules and regulations at the expense of their time in developing the innovative applications. While divulging and occasionally preventing any true conflicts of interest may be warranted, until the academic institutions and government become more open-minded, aspiring innovators must learn these obligations and trudge through these all too often burdensome obstructive requirements and regulations.

The remainder of this article will review the reasons this pharmaphobia narrative is ill founded and explain to potential innovators/inventors (and even health care institutions) how to navigate these regulations.

There are several reasons why understanding this pharmaphobia narrative is important.

First, four facts obscured by this pharmaphobia narrative are important when overseeing this topic.

(1) Many improvements in health care over the last 50 years are a direct result of partnerships between academia, individual clinical doctors, innovators, and industry. Several ready examples include the fact that cardiovascular mortality has decreased by 60%, longevity has increased by 10 years, and quality of life for those with either degenerative arthritis or inflammatory arthritis has improved markedly, not only through the introduction of antiinflammatory medications, but also via the success of joint replacements—all are direct results of close and constructive partnerships with industry.

(2) The tools used by health care professionals, such as antihypertensives, cholesterol-lowering drugs, diabetic medications, the recent introduction of biologics to treat cancer and inflammatory diseases, immunosuppressants for transplantation, endoscopic devices to detect and remove precancers through screening programs, and the ever longer lasting mechanical joints and heart valves have all developed through the partnerships between academe and industry.

(3) Many, perhaps most of these innovations could not have been developed by either academe or by industry alone.

(4) Future development and acquisition of these partnership-developed tools were expensive and difficult to develop, and if the system does not change from its current pharmaphobia, future development and marketing of such innovations will be seriously threatened [5].

A second reason involves the influences of many critics who have tried consciously to minimize the importance of such partnerships between academe and industry. These critics have discounted the advances in health and the improvements in quality of life and well-being related directly to important contributions by industry. Moreover and importantly, they have also overlooked the obligate costs and financial implications of these improvements in quality of life and health and well-being. In doing so, these critics have implied that researchers, health care professionals, and even health care institutions have benefitted financially at the expense of the public and have also implied by their critiques that much of the research carried out is both flawed and harmful to the public. In my book, I have called these critics "conflict of interest narrative instigators," because they use the term "conflict of interest" as an assumption of their accusation.

My book was designed to show that these critics are both wrong and potentially dangerous. These critics are dangerous because of the implications and resultant actions of their assumptions, the subsequent resultant inability to carry out and market innovative approaches and devices, and all the regulations they have initiated lack basis in fact. In my book, I have taken an evidence-based approach using documented facts to counter these false accusations by these conflict of interest instigators.

No one questions the ultimate motive of the truly dedicated physicians and researchers who push the boundaries in medicine to lead to worthy and important advances, but also, one must recognize and acknowledge that the primary engine that ultimately drives innovation comes from industry, usually via the partnerships between practitioners/researchers/innovators and even health care institutions and private industry. My career has been involved with many aspects of research, and I have seen and understand that the development, marketing, and ultimately the availability to the public of innovative approaches and ideas designed to save lives and/or improve quality of life usually requires the expertise and often the funding supplied by medical device and drug companies; this need for partnership has been, is, and continues to be important.

The danger of these conflict of interest instigators is that they distort the facts and use a flawed logic in their pharmaphobic narrative by putting forth supposed "ethical" messages based on past history that is no longer relevant in today's world. These critics compare innovators of today with practitioners in the past who were charlatans pushing patent medicines as cure-alls to make money as a "business" at the expense of society. But we all need to recognize that health care delivery IS A BUSINESS, and we cannot overlook this fact. But to then claim that the devoted physicians, PhDs, innovators, and inventors who accept payment from industry for their hard work and innovative intellectual property are untrustworthy and lack a basic altruism is both ridiculous and dangerous, because this assumption foments the attitude of pharmaphobia which impedes the development of truly innovative ideas.

It is this environment generated by these conflict of interest instigators that has given birth to regulations touted to "protect the best interests of the public." Unfortunately, these regulations have served to impede innovation. Health care institutions have given too much voice under the guise of protecting the public to these conflict of interest instigators by imposing stifling regulations on the investigators trying in good faith to partner with industry; these regulations have been accepted by the press as necessary when in truth, these regulations waste the time of investigators and often dictate arbitrarily and inappropriately what research and partnerships with industry can be explored and pursued. All of this justification of institutional and governmental oversight is done under the implication that the primary investigators and their partnership with industry act unethically to fulfill their own parochial financial benefit at the expense of the public's trust. But when asked to prove this unethical behavior or to show objective proof of financial gain of the hardworking researchers, the proof is lacking, and most all that is left is an attempt to imply a nonexistent smoking gun. And what is evident is that this repressive paranoia of regulation has pervaded widely into all aspects of research, innovation, interactions and partnerships with industry, and ultimately in the current delivery of health care nationwide. The public has been warned and "educated" not to trust physicians, pharma, clinical research trials, and worst of all the basic concept of research and innovation. Focus has centered on such trivial topics as the handing out of pens and refreshments and the sponsoring of professional meetings by industry. The downstream effects include the inability of pharma to provide samples of medications for needy patients and curtailing educational visits by drug representatives to inform busy physicians and health care associates about new drugs or devices. All of this oversight has been established under the unsubstantiated and unwarranted guise of "protecting the public" from greedy physicians, researchers, and industry.

My book explains that a handful of widely publicized incidents within the ever-enlarging enterprise of health care delivery and research billed as examples of industry-related corruption have been misrepresented and magnified immensely by the conflict of interest instigators not only in their institutions but also in the press and the public's eye. Indeed, administrative bureaucrats, aggressive but uninformed reporters in the press, lawyers, and naive administrators have used these accusations to promote their own careers at the expense of hard-working, dedicated researchers. Whether this alleged corruption is real or not, it sells. Politicians get political attention and capital by "protecting the public," lawyers profit financially whether they succeed or not, and newspapers sell. This concerted attack against research, dedicated researchers, clinical trials, and industry partnerships has occurred at the expense of the unwary public by suppressing the marketing of new drugs and devices and thwarting the development of innovative ideas.

Who are the parties that were responsible for this environment of regulatory control? Alas, the instigators have arisen from our own ranks—the bureaucrats within our institutions. Indeed, these internal executives of our health care system continue to pay lip service to the need and desire for partnerships with industry, but respond with an unfounded paranoia to the allegations of the conflict of interest instigators by establishing the burdensome rules designed ostensibly to prevent corruption but which actually suppress research and innovation; these policies are all done to avoid embarrassment to their careers and to feed the appetite of the instigators to avoid any bad press, whether it be true or not. This concern of conflict of interest has invaded the health care literature as well. Just witness the attestations required of medical authors when submitting their work. While the full and transparent documentation of conflict of interest and open acknowledgment of industry financial support may be justifiable in the literature, the knee jerk response to industry support by reviewers, journal editors, and readers has been one of overriding concern and mistrust of the data reported. Indeed, the fallout of this concerted attack by the conflict of interest instigators has been that one is guilty until proven innocent.

Because of such unsubstantiated and inappropriate maligning of industry and investigators (i.e., guilty until proven innocent), the development of innovation in health care has been impeded seriously by the conflict of interest movement by both the press and our own institutions who are more worried about bad press (be it real or not!). The fallout has been that the required compliance with and enforcement of all the regulatory attestational regulations has only served to divert both finances and resources away from meaningful research and partnership with appropriate interested industry.

My book describes three specific examples of how pharmaphobia results in inhibition of innovation. One is the results of a survey of inventors having established or attempting to establish industry partnerships that revealed a substantial number of developmental projects being delayed or prevented due to conflict of interest concerns. Given the very low success rate of development of health care products, any delays or derailments of projects are unacceptable.

A second example is that prosecutors have employed the rhetoric of pharmaphobia to abuse "false claims" litigation. Conviction from this type of legal strategy triggers a draconian penalty—companies cannot sell products to government payers. The lethality of this outcome forces companies to settle almost certainly unprovable corruption claims. The settlements incur enormous—$ billion—fines and saddle companies with onerous compliance burdens. Even the threat of this type of legal predation causes companies to curtail interactions with physicians and inventors.

The third is a component of the controversial affordable care legislation known as "The Physician Payments Sunshine Act." This law mandates that health care product manufacturers report to the Center for Medicare and Medicaid Services for pubic posting any payments to licensed health care practitioners in cash or kind exceeding $10.00. Since implementation in 2014, the reported data have revealed that the vast predominance of payments are trivial and that larger amounts are all for legitimate research interactions. Early on, muckraking reporters used the information to shame providers receiving large amounts, but otherwise, nobody has paid attention to the database. In fact, providers, especially those not in academic settings, are not even aware of the law, although when apprised of it, profess to its distancing them from industry interactions despite believing that such segregation will impair patient care [6]. This useless and expensive exercise is hardly compatible with innovation.

The argument that drug and device development is not extremely expensive is a myth and encourages taxing of companies and research endeavors or imposition of price controls, both of which compromise innovation. These effects have and continue to delay the rolling out of true innovations, which will occur at the expense of patients with disorders requiring these innovative treatments that offer a potential cure or improvement in their quality of life. The prevention of educational, informational interactions between physicians and other health care workers with industry representatives serves to delay the recognition and adoption of new treatments. Similarly, the unnecessary and burdensome regulations imposed on the research environment (both by health care institutions and by industry) that delay potentially innovation-promoting interactions, sponsorship, and true partnerships have been documented but are all too often overlooked and denied by these conflict of interest instigators to justify their role in protecting the public from the "greedy doctors." This paranoia is pervasive and present in the government, health care administrative offices, and unfortunately in the public realm.

REFERENCES

[1] Zycher B, DiMasi J, Milne C. Private sector contributions to pharmaceutical science: thirty-five summary case histories. Am J Ther 2010;17:101–20.

[2] Sampat B, Lichtenberg F. What are the respective roles of the public and private sectors in pharmaceutical innovation? Health Aff 2011;30:332–9.

[3] Stossel T. Removing barriers to medical innovation. National Aff 2017;68–82.

[4] Stossel T. Pharmaphobia: how the conflict of interest myth undermines American medical innovation. Lanham, MD/Boulder, CO/New York, London: Rowman and Littlefield Publishers; 2015.

[5] DiMasi J, Grabowski H, Hansen R. Innovation in the pharmaceutical industry: new estimates of R&D costs. J Health Econ 2016;47:20–33.

[6] Barton D, Stossel TP. Pediatric dentists' knowledge concerning the physician payments sunshine act and their predictions of its effects on their interactions with the industry and its impact on patient care. Pediatr Dent 2016;38:122–6.

Chapter 15

Adoption of Technology: Appealing to the Hospital and Health System Value Analysis☆

Jenell Paul-Robinson
MedTech Innovator, Los Angeles, CA, United States

The genesis of value analysis began in the mid-1990s with the impending balanced budget act of 1997. Behind the number one operational cost of labor, most organizations quickly understood that supply costs are the second largest component of overall spending. This realization forced hospital organizations to pay close attention to what comes in, how products come in, and exactly how products are being used. In order to reduce costs and ensure high medical quality, health care institutions formed committees to reach their provider level leadership, physicians, and nurses. Committees were formed not only to cut costs, but to make impactful decisions regarding individual organizational missions. Fast forward to today, the same principles apply but on a much grander scale. Changes in the reimbursement climate have forced value analysis committees to think differently and well beyond supply expenditures, due to the low hanging fruit or conversion to less expensive like-items being exhausted. Organizations are now looking outside the box to ensure that the products and technologies that pass through their doors create a much larger impact—the impact to ensure quality care and improve patient outcomes, as well as to save money in the context of total cost.

What exactly is value and how is it assessed? This depends on who is asked. Value analysis is an umbrella term aimed at assessing improvement to the patient, payer, provider, and health care ecosystem. While each organization

☆ *Disclaimer*: This is purely an observational opinion piece. It is not bound by published literature or documented data. It is solely a report on my experiences, both locally and nationally, through my own industry experience as a value analysis coordinator for a world-renowned academic medical system.

Medical Innovation. https://doi.org/10.1016/B978-0-12-814926-3.00015-2

assesses value differently, the overarching goals remain constant, aimed to support patient-centric, quality, outcome-driven treatment options while decreasing the overall cost of care. Today's supply chain decision-making is complex and incorporates disciplined and concrete processes to assure the products utilized are appropriate and have a positive impact. In the current climate of reimbursement, the stated common-goal factors are the foundation of the organizational process of American health care decision-making.

To address the ever-changing health care system, hospitals formed and empowered value analysis committees which are charged to perform due diligence, ensure product purchases are quality-driven, fiscally responsible and provide clear outcome advantages in an environment of patient-centric care. Innovative ideas and products that drive positive outcomes are welcomed, creating many opportunities for the early start-up; however, value must be demonstrated in order for the product or service to have a fair chance of being added to the hospital's inventory.

The shift to value-based care has affected decision-making on the macro level. All products coming into health care organizations are now scrutinized heavily for their contribution to the value equation. Long gone are the days of "nice to have." In the current climate, products must clearly address an unmet need while providing an advantageous edge to current or standard practice or patient outcomes. The future of supply chain management will depend on justification as to the products and supplies utilized. The need request must clearly fill a void or improve outcomes. It is imperative for innovators and early start-ups to articulate clear benefits above and beyond elementary cost-savings, especially in the space where strong clinical evidence may not be readily available.

Currently, there is no standard or magical formula on how to assess value. Leaders in the value analysis space, however, are emerging. Consistent processes, administrative support, committee structure, and data-driven decision-making are the common threads among the high achievers.

The most effective value analysis committees take a holistic approach to both the process of structure and decision-making. High achieving, value analysis committees tend to be structured with representation from multidisciplinary services. Key components of successful value analysis programs may include a centralized decision-making, peer-reviewed process consisting of representation by hospital senior administration, general medicine, infectious disease, varied surgical services, anesthesia, nursing, supply chain, managed care, risk-management, and financial decision support. The diversity of the committee offers multiple angles of insight and expertise. The high level of expertise adds to the rigorous scrutiny of products considered for utilization by the organization.

EVIDENTIARY SUPPORT PACKAGES "SHOW ME THE NUMBERS"

Not all evidence is created equal. In order to provide best-in class care, organizations seek cutting-edge, state-of-the-art technology to provide the most relevant, novel treatment modalities for their community. It can be extremely challenging for innovative start-up companies, even with great ideas, to provide the level of data required for the decision-making process. Product or device approval tends to rely heavily on evidence-based studies of Level 1 or 2, statistically significant, randomized controlled studies and peer-reviewed literature that support improved outcomes. While committees rely heavily on peer-reviewed, published data, it is understood that early start-ups may not have the breadth and depth of desired study data. In this case, efficacy and superiority data are welcomed. Case reports are a good start but probably will not be persuasive enough due to individual patient variables and a strategically selected, optimal patient population. Outcome data or theory validity in as large a sampling size as possible may help demonstrate advantages over existing standards of care.

Careful consideration for data metrics is extremely important. When possible, comparison to the existing standard of care and how implementation of the product improves outcomes is highly desirable while removing as many variables as possible. While passion for innovation is commendable, failing to demonstrate comparative or competitive advantage will not hold water in today's multifaceted, complex, decision-making process. The innovative product may offer a novel treatment approach, but the truth is, the intended patient population is treated currently in some way, and to project a true financial picture and demonstration of measurable outcomes, the financial projection must include the current standard of care for the intended patient population.

Early value propositions tend to focus around cost-savings. While the concept of cost-savings is advantageous, the real value in innovation is multifaceted. Strong value propositions address clinical impact, nonclinical impact, fiscal impact (patient, provider, and payer), and the impact on the health management of the population to be treated. Organizations will identify the intended population to analyze the book of business, payer mix, and outcomes associated with the current standard of care. Innovators are encouraged to define the current standard of care and position their data metrics to compare as closely as possible. While each organization is likely to generate its own baseline metrics, a thorough and honest approach adds credibility for the innovation.

Practical, quantitative metrics are quite easy to capture to provide evidentiary support. Hypothetic and highly variable metrics take additional due diligence. Proposed time-savings is a good example. The concept of time-savings is certainly of interest to hospitals, but overestimation and unsupported justification of product cost is a common mistake innovators tend to incorporate into their value proposition and business plan. Most operating room costs are fixed (labor, etc.); in contrast, supply costs are variable. Equating cost-savings to decreases in operating room time, offset by the cost of the device, does not necessarily hold its weight with value analysis committees. And this is why: Projections of time-savings are highly dependent on multiple variables and typically considered to be a soft savings for committee decision-making purposes. Organizations rely on full operating room schedules to maximize fixed costs.

The question is never "how," but rather "how much?" How much time can be saved in relation to the total procedure time? Does the proposed product produce enough efficiencies and time reduction to schedule another case and maximize fixed labor costs? Remember, 5 min of savings in a 4-h procedure is not nearly as impactful as 5 min in a 30-min procedure. Downstream patient benefits of decreases in operating room exposure are of absolute consideration; however, important questions remain, how much time-savings and what exactly does that look like for both the patient and the organization? The same guiding principles will need to be considered when innovators present value propositions for a proposed decrease in readmission rates.

REIMBURSEMENT: TO BUY OR NOT TO BUY

The decision tree is complex. Many times with emerging technology, organizations are forced to make a decision on a loosely supported reimbursement leap of faith and the obligation to "do the right thing." In reality, reimbursement for emerging technology is highly unknown at the time of review and decision-making. The proposed improvement in outcomes makes sense logically, but the improved outcomes are hypothetical and supported only by conjecture. That is okay and should not deter the start-up from pressing onward.

It may be unrealistic in today's climate to assume hospitals will receive additional reimbursement for existing procedures. Reimbursement is merely a factor in the decision-making process. With the shift to value-based purchasing, considerations concerning episodes of care and outcomes are more important than ever before. The truth is, uncertainties remain in the future of reimbursement, prompting health care organizations to make difficult, complex decisions based on risk vs. reward; this process requires extremely comprehensive analytics to support justification for use. The large number of unknowns causes institutions to band together when the business of health care does not make justifiable sense. For the greater good of the progression and improvement in patient care, hospitals must be willing to take a leap of faith, placing most of the eggs in the proverbial basket of future reimbursement. Take robotics or laparoscopic surgery, for example. If economic decision-making were the sole means of procurement, the approval process may have looked much different. Millions of patients may not be receiving the multiple benefits of less invasive surgery over open operations. The decision to take a leap of faith based on providing a quality treatment modality for the patient was a moral and clear, correct choice—a decision not based on guaranteed coverage or reimbursement.

We are now in a decision-making situation of need and population health management. The combination of morality, quality, and business is certain to become even more intertwined as the reimbursement landscape continues to change. Clearly demonstrated, improved outcomes will move the initiative along in organizations that are willing to take a risk outside the certainties of current coverage and reimbursement.

Contrary to popular belief, reimbursement is not always a sure bet for addition of various products or devices to the medical supply formulary. While the committee decision is much easier, we need to stress that reimbursement status is not the sole decision-making factor. Patient-centered decision-making focuses on outcomes, evidence, existing treatment modalities, gaps in community resources, and finally, finances. The "value" of the proposed product is of utmost importance. If the proposed innovation does not provide clear value and beneficial outcomes to the patient, it is unlikely to be given consideration for addition to the medical supply formulary, regardless of coverage or reimbursement status.

YOU HAVE DEFINED YOUR VALUE, NOW WHAT?

There is an overwhelming number of vendors approaching health systems today. Each is competing for the same time slot on a packed appointment calendar which causes frustration for everyone involved. This oftentimes leads vendors to reach out to hospital administrators (CEO, COO, CMO, and the like). Unfortunately, hospital administrators are often the wrong target audience to promote the purchase of the product. While the "C-Suite" is an integral cog in the decision-making wheel, the structure of most value analysis committees is intended to engage clinicians and key opinion leaders, enabling a holistic approach to decision-making. The correct audience is the clinician—it is the clinician who will care for the patient, use the product, follow the patient through their journey, and have a vested interest in associated outcomes. This is your clinical champion. The clinical champion is the individual who is passionate enough about your product to complete a limited trial for usefulness, shepherd the product through the organizational process, or perhaps partner for publication.

In order to navigate the complex value analysis structure, it is of particular benefit to understand the intricacies of key organizations. Value analysis professionals are a rich resource to truly understand organizational needs, processes, and key opinion leaders.

While there are many similarities among the community of health care organizations, many differences exist. It is important to fully understand how each organization operates, respecting their processes, and realizing that no two are exactly alike.

While often used synonymously, it is important to point out that value analysis processes are not purchasing processes and vice versa. While the two processes are linked, they are not one and the same; each has their own procedures and cadence. Value analysis is made up of multidimensional decision factors. Purchasing is the transactional process a product goes through after the decision to purchase is made.

Each organization is different as to the thresholds for departmental approval concerning procurement processes and authority. The same holds true for expenditures for capitol equipment. Holding discussions regarding purchasing and procurement limitations and processes with the local value analysis or strategic sourcing coordinator will help clarify expectations. Depending on the organizational structure, value analysis processes may not have authority over decisions of associated capital equipment purchasing. The innovators must vet out each organization separately and avoid the assumption that the organization across the street operates in the same fashion.

COMMITTEE EXPECTATIONS

Preparation is key in order to move forward and introduce a product. In-depth homework upfront will save time and angst. Due to the complexity of the innovation request, one can understand why there is no standard crystal ball for the time frame of the request to decision. Each hospital will assess need differently. While the overarching factors are largely the same for all hospitals, the priority across different institutions and hospitals varies (at times even within institutions under the same organizational structure). Priority variations may include capital equipment cycles, strategic programs, or service line growth objectives to meet community needs. This may help describe why an item can be approved for use at one hospital but is not approved at a sister facility.

The product request is likely to be presented before a room of experts across multiple disciplines and disease modalities. Validating the innovation with experts in industry is integral to providing relevant information for presentation. Successful tactics include utilizing a variety of industry experts across multiple geographic regions. This strategy will cast the widest net of opinions possible, potentially decreasing geographic and other biases. Proper validation of the product will diminish the risk of compromising credibility.

Based on the sensitive information circulated within the committee, it is common for meetings of the value analysis committees to be held behind closed doors. This process is not intended to be a deterrent or secretive, it is merely due to the confidential information concerning finances and contracting discussed when analyzing impact of the proposal. Most innovators are able to navigate this by working with the value analysis coordinator and physician champion to provide the committee with a clear value package for presentation.

Value analysis committees can make varying degrees of approval recommendations. Aside from unconditional approval to the medical supply formulary, the committee may consider variations of conditional approval pending additional information, or denial. HOW the request is presented can be just as persuasive as WHAT is being requested.

Should the initial decision be unfavorable, do not give up. It is fair to ask for specifics of the denial. Unfavorable decisions can be amended, allowing the clinical champion to present the most compelling, outcome-driven, cost-effective story for reconsideration.

While organizations may not always agree to purchase new technologies at first review, interests align with the intentions of the start-up. Innovation is critical to improving patient outcomes. Systems are willing to pay for technology that delivers better outcomes, but the value must be stated clearly and with a good argument for benefit to the patient and/or institution. Promising technologies will be given another chance to return to the committee for additional consideration. Stay positive. Be persistent. Timing, internal clinical support, and a solid value proposition are key. Be prepared to pivot, make amendments, and keep forging forward to make the world a better place.

PREPARE YOURSELF APPROPRIATELY

Today's value analysis decision-making processes are highly complex. Start-ups armed with a few basic key considerations can greatly improve the adoptability of emerging technology.

- Be honest and specific in your approach
- Solid value propositions are patient-centered
- Fully understand your disease process, the patient journey, and where the application of your product will have the greatest impact
- Avoid the common mistake of stating that the product is void of competition. When in doubt, study the current standard of care to determine where the proposed product fits in the process
- Understand and respect the differences and nuances of individual processes of hospital organization
- Recruit a clinical champion who is passionate and willing to shepherd the product through the process
- Understand that not all evidence is created equal. Study the evidence hierarchy structure to determine the kind of study that creates the greatest evidentiary support for your proposed outcome claims
- Collaborate with the value analysis professionals; they are a rich resource.
- Global financial projections will be heavily scrutinized and reworked (over and over). Do not be offended. Hospitals must match metrics applicable to their specific patient population
- Understand the difference between coding (physician and facility), coverage, and reimbursement
- Be prepared to exercise patience

Chapter 16

Technology Adoption: Appealing to Payers and Capturing Economic Value

Carla L. Zema
Vice President, Precision Health Economics, Los Angeles, CA, United States

INTRODUCTION

You have a great idea for a new invention. You have done your research and know that the invention has the potential to revolutionize the current way patient care is provided. In addition, you have planned for exactly what is required to obtain clearance by the Food and Drug Administration (FDA) and believe you are all set to move forward. Historically, this might have been all that was necessary to move a great idea into a viable invention and to drive your concept to commercialization. With the changing landscape of health care; however, this is no longer the case. Moving a good idea through to regulatory approval is insufficient to ensuring the invention reaches patients. Inventors must also consider what is necessary to ensure that the invention gets adopted after regulatory approval. Unfortunately, there are many skeletons of great inventions that achieved regulatory clearance but were never adopted.

Central to adoption is reimbursement. Therefore, this chapter will present considerations for getting your invention reimbursed. A broad definition of reimbursement will be taken, because it means different things depending on the type of device. For the purposes of this chapter, the term reimbursement will be inclusive of payments, purchasing, coverage, and tendering. Reimbursement pathways including evidence requirements will be explored along with a discussion of provider-based, risk-sharing agreements, also commonly referred to as value-based purchasing. Although this chapter will focus on the health care market in the United States, considerations for future global reimbursement are important and will be discussed briefly. The goal of this chapter is to provide practical knowledge and recommendations for how reimbursement should influence the early development process.

Medical Innovation. https://doi.org/10.1016/B978-0-12-814926-3.00016-4

AFTER REGULATORY APPROVAL: GETTING THE DEVICE TO PATIENTS

In the current health care landscape, reimbursement is necessary in most instances in order for your medical device to reach patient care in the clinic, operating room, hospital, or other health care setting. Reimbursement can mean many things; an oversimplified way to think about reimbursement is who will pay for your device, how they will do it, and how much will they pay. Just as there are various pathways to regulatory approval or clearance, there are many different pathways to reimbursement that vary depending on the type of device. Examples of different types of medical devices include capital equipment (e.g., MRI, surgical robot), medical supplies (e.g., syringes, IV tubing), implantables (e.g., joint replacements, pacemakers), and durable medical equipment (e.g., wheelchairs, wound vac). Of course, there are other categories as well as devices that can cross multiple categories, such as software for medical informatics, various forms of the electronic medical record, novel administrative initiatives, diagnostic tests, organizational paradigms, etc. Just as complicated as attempting to categorize devices, the nuances of reimbursement can be quite complex. An overview of the more common reimbursement pathways will be discussed.

Reimbursement Pathways

Although there are many pathways to reimbursement, many health care stakeholders think about reimbursement as the process by which a health plan determines what types of products or services it will reimburse or the level of reimbursement it will provide. Consider a newly approved drug that gets prescribed directly to patients. After FDA approval, health plans will review the information and the objective evidence to decide whether to put the drug on its formulary which then will determine the level of reimbursement to patients when they are prescribed the drug. This pathway in which a health plan determines whether or not to pay for a product or service is what is most often thought about when considering "reimbursement." In fact, much of the economic literature, industry-based reimbursement roadmaps, and policy discussions around the appropriate reimbursement for a medical device focuses on this reimbursement pathway. Ironically, very few devices go through this pathway. Let us start with a contextual foundation for reimbursement.

You may have heard the term health technology assessment (HTA), which is the systematic evaluation of the impact of health technologies usually for purposes of adoption of these new technologies. HTA examines the clinical effectiveness and safety together with the intended and unintended consequences of the technology (and sometimes cost), to determine the impact of the adoption of the technology on the health care system. HTA is intended to evaluate the value of health technology; hence, *value assessment* is often used interchangeably with HTA. This process can be formal, such as in many countries outside the United States like Great Britain, or informal as with many US health plans. While informal assessments may not be named "HTA" or "value assessment," the review process is virtually the same.

Other terms that you may hear are comparative effectiveness, cost-effectiveness, cost-benefit, etc. Comparative effectiveness is a broad term that encompasses many specific types of comparative analyses and is equivalent to HTA. Cost-effectiveness is a specific type of comparative effectiveness analysis that includes costs. The 2010 Patient Protection and Affordable Care Act created the Patient-Centered Outcomes Research Institute to conduct research in comparative effectiveness that could potentially inform both public and private health care decision-making. This funding came with the specific provision that the organization could not consider costs in its research. Despite this provision, however, most stakeholders agree that the inclusion of costs is critical to informed decision-making.

Currently, there is no standardized format for how information and evidence should be compiled and presented. For pharmaceuticals, most decision makers in the United States follow the format established by the Academy of Managed Care Pharmacy (AMCP). While the AMCP states that their format can be used for devices and diagnostics, and several other organizations suggest formats as well, there is no clear standard or requirements for medical devices and diagnostics. Despite the lack of a standard, there are some best practices when preparing information and evidence during review for reimbursement.

Description of the *Disease or condition*: What is the epidemiology, what are relevant risk factors, and which are the relevant subpopulations? What is the associated pathophysiology, clinical presentation, and burden of the disease?

Approaches to treatment: What are the current standards of care, what are the relevant options, and where is the appropriate setting of care? Guidelines and professional consensus statements are helpful.

Description of the Device or new technology: What is the device or technology being considered? How it is used? Who uses it? And in what setting?

Your device's place in the current treatment: How will your device fit into the current standard of care? It is intended to add incremental value and to what extent or will it completely replace current treatment?

Clinical effectiveness: What are the clinical benefits of your device, and what is the evidence to support your contentions?

Economic benefit: What is the realistic economic impact of your device?

This is an extremely high-level overview of the key elements that will be considered during an assessment. Of course, there are numerous considerations in the details of this information and evidence. While some additional detail will be provided in this chapter, appropriate information to be included varies by type of device.

A key component of this process is a comparison with competitors of the technology being reviewed. This comparison may include the obvious competitors, such as current devices in the same class of treatment or alternatives, such as radiation, chemotherapy, or surgery for prostate cancer. Choosing the right comparators is critical in this process, because different comparators can change the results of the assessment. It is quite possible that the organization conducting the review might choose different comparators than what you would consider. Therefore, it is important that the information and evidence that is pulled together for reimbursement purposes have clear references and justifications for comparators as well as for the therapies that should not be compared.

Public and private payers have reimbursement processes, often referred to as medical policies, for accepting and reviewing this information to determine whether the device or service will be reimbursed. An example of a policy for cochlear implants in a commercial health plan can be found at https://www.uhcprovider.com/content/dam/provider/docs/public/policies/comm-medical-drug/cochlear-implants.pdf. The Center for Medicare and Medicaid Services (CMS) makes determinations of national and local coverage regarding such policies for Medicare and maintains a database located at https://www.cms.gov/medicare-coverage-database/overview-and-quick-search.aspx?list_type=ncd. Many state Medicaid programs also have online databases available to find information on specific medical policies. The reality is that very few medical devices go through this pathway for reimbursement. Types of devices that typically would go through this pathway are more commonly the durable medical equipment (referred to as DME) and devices used directly by patients, usually at home.

The vast majority of devices, however, are used in hospitals and in other health care settings by health care providers (HCPs) and often times as part of a broader procedure. Another reimbursement pathway is procurement or direct purchasing. Consider a hospital that purchases medical devices and products for use in their facility. Some of these products, typically the undifferentiated medical supplies, go through a purchasing process. More relevant to inventors is the purchasing process for most of the medical devices used by hospitals. This type of process usually involves submission of similar information as described above and is to Value Analysis Committees (VACs) for review and consideration of adoption within the facility. This process is discussed in greater detail in Chapter 15.

Coding

In preparing for reimbursement, other activities related to medical coding are needed. Medical coding is the mechanism that supports reimbursement both directly and indirectly, depending on the device being considered. There are several types of standardized coding schemes that are relevant for medical devices. Most procedures have an associated Current Procedural Terminology (CPT) code. These five digit numeric codes are maintained by the American Medical Association and were last updated in 2017. Some procedures also have an International Classification of Diseases, 10th edition (ICD-10) code. While many are familiar with ICD-10, Clinical Modification (ICD-10-CM) diagnosis codes, ICD-10 also has less well-known Procedure Coding System (ICD-10-PCS) codes which are not used as widely as the CPT codes. In the United States, one of the most important coding systems is the diagnosis-related group (DRG), which is used to classify hospital cases into one of approximately 500 groups based on similar expected uses of resources. DRGs are the driver of the Medicare Inpatient Prospective Payment System (IPPS) used by CMS. DRGs are assigned using a "grouper" program based on ICD-10 diagnoses, procedures, demographics, discharge status, complications, and underlying comorbidities of the patient. Durable medical equipment, prosthetics, orthotics, and supplies (DME-POS) that are used outside of hospitals and physician offices are captured typically by the Health care Common Procedure Coding System (HCPCS) codes. These codes begin with a single letter followed by four numeric digits. Understanding which of these coding systems is most relevant for your device and the process for either getting a new code assigned or having your technology added to an existing code is vital for reimbursement.

Think about how your device should be coded if it is to be used. Is it part of a procedure? An example would be a prosthetic knee joint used in a total knee replacement. There are ICD-10-PCS and CPT codes that feed into a DRG. How much does your new technology change the procedure? In other words, should your invention be considered an improvement but part of existing coding, or is new coding required? Generally, resource utilization is an important consideration for the respective committee in this decision. An example is the leadless pacemaker. Traditional pacemaker systems are implantable cardiac devices that consist of a generator that gets implanted in a subcutaneous "pocket" in the chest wall with electrical "pacing" leads that are placed directly in the chambers of the heart and adherent to the wall on the heart chamber. The FDA approved this leadless pacemaker recently, which paces the heart without the need for the leads of the classic pacemaker, a major advance in cardiology. The procedure codes for insertion of a pacemaker system include the operative insertion of the generator and the leads of a traditional pacing system. Should the leadless pacemaker be included in this existing code or should a new procedure code be created? One of the initial innovators of the leadless pacemaker applied for new procedure codes. Temporary CPT codes were issued (these codes typically end in a "T" while the standard CPT codes are only numeric). Experience over several years will be evaluated to determine final changes to the CPT coding system for leadless pacemakers. These new CPT codes also trigger evaluation of how these new codes feed into existing DRGs. There are currently three DRGs relevant for pacemaker insertion: (1) pacemaker insertion without complication or comorbidity, (2) pacemaker insertion with complication or comorbidity, and (3) pacemaker insertion with major complication or comorbidity. Similarly, evaluation of the impact of the new CPT codes on the related DRGs will be monitored over the next several years with potential modifications considered based on the results.

The process for obtaining new codes and evaluating their impact on reimbursement takes several years. Inventors may have never considered this as part of the development process. But, think about this from a hospital's perspective. You have developed a new implantable technology that revolutionizes a current procedure and has substantially decreased the likelihood of future complications and device replacement, and the new device is quite a bit more expensive than the current technology, but the procedure reimbursement will remain the same for the hospital. Hospitals face these situations often when considering new technology. Budgetary pressures force hospitals to make tough decisions on the best use of their limited resources. Therefore, reimbursement is critical to adoption decisions.

Payer

As illustrated in the reimbursement pathways described above, the "payer" in various situations can be different. It is important that you identify not only who the payer will be for your device but also other relevant stakeholders in the reimbursement process. Ultimately, you need to know who is financially responsible when your device is used. This question is not as easy as it may seem. In some instances, it may be a health plan that reimburses for the use of the device. Does the patient have any responsibility? For example, many benefit structures of heath insurance plans now have separate deductibles for DME. Consequently, even if the health insurer decides to reimburse for your device, a patient with a $10,000 deductible for DME that is separate from his or her deductible for medical services means that the patient may likely have to pay out of pocket for a large portion of the use of the device. Another example might be that the device gets reimbursed by health insurance, but the physician's office must stock the device in their office. In some cases, physicians keep the products in their office and bill the health insurance once a patient uses the device. An example of this would be heart monitors. In other situations, the physician actually purchases the device but only receives reimbursement from a health plan when the device is actually used by the patient. In this instance, the inventor must ensure that both the health plan and the physician are considered in the planning for reimbursement during the considerations needed before embarking on the expenses and effort of developing the device.

Hospitals are often the payer as in the example of a knee replacement or even direct procurement of many medical supplies. With a knee replacement, health insurers are also stakeholders, because they pay for the procedure itself. As discussed in Chapter 15, hospitals have a process for evaluating and adopting new technology; this process may or may not involve a formal VAC, but it will certainly involve input from a physician(s) and administrator. One must consider not only the physicians who will use your device directly but also the ones who will be affected if the standard of care changes. When cardiac stenting was developed, the standard of care was open heart coronary artery bypass operations. Stenting allowed cardiologists to intervene with a much less invasive, percutaneous transvascular approach, thereby decreasing the number of open heart operations performed by cardiothoracic surgeons (and for the hospital). Similar cases occur with the current innovations in value replacements, where traditional value replacements were performed as open heart operations. These newer valve replacements can often be inserted transvascularly through a catheter placed percutaneously by an interventional cardiologist/cardiac surgeon. Initial resistance from cardiothoracic surgeons and hospital administrators based on the powerful financial incentives of the original open heart operations probably delayed the adoption of the stenting longer than necessary when considering the benefits to patients alone. One must not be naïve and remember that reimbursement and financial incentives are powerful drivers of decisions regarding adoption of new devices/technology, and these financial drivers should not be underestimated.

EVIDENCE REQUIREMENTS FOR REIMBURSEMENT: NEED FOR EVIDENCE-BASED DATA

You probably already know the evidence requirements (evidence-based data) for regulatory clearance or approval. Unfortunately, the evidence requirements for reimbursement are typically not aligned with the evidence requirements for regulatory clearance or approval. The evidence required through the federal regulatory process is determined by the regulatory pathway. The evidence required for a 510(k) application is less than for an application for a premarket approval (PMA) given the nature of the application processes. In a 510(k) application, regulators are looking for evidence that your device is equivalent to a predicate device. You do not necessarily need to provide evidence for clinical effectiveness and safety as long as you can show that your device is equivalent to the predicate device. There are far more 510(k) applications each year than PMA applications.

The level of evidence required to support reimbursement generally follows the traditional evidence hierarchy in which double-blinded, randomized clinical trials are the gold standard. Thus, a major gap exists between the evidence required for regulatory clearance and for reimbursement. In fact, most of the examples of devices that achieved regulatory clearance but were never adopted suffered from lack of sufficient evidence for reimbursement which resulted in strong financial incentives not to adopt the new technology. In fact, many current investors are well-attuned to this issue and expect considerations for reimbursement and adoption to be part of your development plan and a necessary condition for funding. Both regulators and payers acknowledge this gap and continue to work together to address this dilemma.

What does this mean for your strategy of clinical development? Given this evidence gap between regulatory and reimbursement, savvy inventors must consider the necessary requirements of reimbursement evidence when developing their strategy for generating clinical evidence. Just as you may have gotten input from regulatory officials on the evidence required for clearance, the inventor would be wise to also seek input from reimbursement stakeholders and experts. This investment of time and effort may result in changes to your current study or the need for additional studies that will be required for reimbursement. Waiting until you complete your initial clinical studies will result in delays in adoption and will most likely mean wasted resources to produce evidence through additional studies that might have been able to have been generated through modifications to current studies.

One challenge with generating evidence for reimbursement is that payers often have unrealistic expectations of the level of evidence possible. Double-blinded, randomized controlled trials are not always feasible for devices, because blinding of the physician and the patients involved in the study may not be possible nor even ethical. Large randomized trials may not be feasible financially for small, start-up companies and entrepreneurs who lack the capital to invest in such trials or the access to the appropriate patient population. The unintended consequence is the loss of a potentially important innovation.

Another challenge that faces inventors with regard to reimbursement is the chicken-or-egg dilemma of user demand vs reimbursement. Guarantee of appropriate reimbursement certainly provides a positive financial incentive to increase use of the device, but payers will often ask for user demand as evidence of need. During the collection of supportive clinical data, consider collecting supplemental information on how the device can best be integrated into regular practice. The strict, well-controlled design of many clinical studies and the narrow inclusion criteria involved in patient selection enforced to maximize the desired outcome can often be different than "real-world" use, and it will be important for you to understand what can be done to support implementation. This is also a great way to involve potential users who may not be part of the clinical study in the development process.

PROVIDER-BASED, RISK-SHARING AGREEMENTS (PBRSAs)

A chapter focusing on reimbursement would not be complete without a discussion of PBRSAs also commonly referred to as value-based purchasing (VBP), value-based contracting (VBC), and outcomes-based contracting. Generally, there are two broad categories of PBRSAs: coverage during evidence generation and reductions in effective price [1]. Most stakeholders, however, and especially payers, think of price modifications through contracting when defining PBRSAs. These types of arrangements are increasing rapidly and are motivated by payers, providers, and manufacturers.

Coverage during evidence generation is a type of PBRSA in which a payer agrees to reimburse for the use of the device while clinical evidence is being generated; this approach is a potential solution to the dilemma discussed previously of reimbursement vs adoption. CMS has a formal coverage program during evidence generation, referred to as Coverage with Evidence Development (CED), for the Medicare program. While certainly promising, CED is quite rare. Since the program's inception in 2005, CMS has used the CED program in only 19 cases [2]. In November 2011, CMS decided to begin to revamp the process and issued new guidance in 2013. The leadless pacemaker is an example of a technology in the CED program. Additional information on this example can be found in the National Coverage Decision for leadless pacemakers available at: https://www.cms.gov/medicare-coverage-database/details/ncd-details.aspx?NCDId=370&ncdver=1&NCAId=285&bc=ACAAAAAACAAAAA%3d%3d&.

The other and more common type of PBRSA, reductions in effective price, can be at the patient or population level. They can be simple such as volume-based pricing or more complex, such as outcomes-based pricing or rebates. While outcomes-based pricing is the stated goal of many stakeholders, there are many challenges with implementing such agreements. Determining the appropriate measure of outcomes can be difficult, confusing, and frustrating. Even when the right metric can be determined, data and numerous issues with implementation can be daunting and may prevent implementation of such agreements. While you may not be involved or even aware of future negotiations for PBRSAs, it is helpful to consider the downstream possibilities. Early clinical studies may set the stage for how future agreements need to be designed. While this is certainly a lesser priority when designing development studies, it is helpful to at least consider future implications.

GLOBAL CONSIDERATIONS

While this chapter focuses primarily on the United States, it is important to at least consider future plans for global dissemination, especially when it comes to the generation of evidence for purposes of reimbursement. Most countries outside the United States represent single-payer systems with more formal processes for HTA and reimbursement. Procurement also is typically more formal. Single payer systems can sometimes issue tenders for purchasing and acquisition of devices. Other devices may be reimbursed through being included on a formal list of allowed medical supplies. Regardless of the pathway, the level of evidence required is greater than is needed for regulatory approval and has been perceived as more difficult than in the United States for many countries. In terms of regulatory approval, however, the evidence requirements outside the United States have historically been perceived as less than for FDA clearance or approval. In fact, many device inventors have chosen to seek regulatory approval outside the United States before going to the FDA. The CE Mark is the regulatory requirement for commercialization in Europe and several other countries. Obtaining a CE Mark by any of numerous "notified bodies" throughout Europe allows commercialization of your device in every country that accepts the CE Mark. In 2017, the European Union adopted new Medical Device Regulations (MDR) and In-Vitro Diagnostic Regulations

(IVDR). These regulations are intended to strengthen premarket evidence requirements. It is not yet known how these requirements will compare to FDA standards in the US or to those in other countries. Like in the United States, changes are occurring in the regulatory process in Europe to close the gap between regulatory and reimbursement evidence requirements.

CONCLUSIONS

Thinking about reimbursement and real-world adoption after regulatory clearance or approval is probably a new concept for many inventors. Having your invention reach patients through broad adoption is probably the best measure of success of your invention. Reimbursement is, unfortunately, a major driver of adoption and must be considered early in the development process, even before the strategy for generation of the needed clinical evidence is finalized. There are many details and nuances to reimbursement that vary depending on the type of device, so it is recommended strongly that you consult with reimbursement experts and your future payers during the development strategy for evidence generation. Consider not only who the "winners" are with potential adoption of your device but also who the "losers" are going to be. Bringing both sets of stakeholders into the development process early can only help to support adoption after you achieve regulatory clearance or approval.

IMPORTANT POINTS

- Central to the adoption of your invention/device/concept is reimbursement, remember, reimbursement means different things to different people and may be different for different innovations
- Develop a reimbursement strategy BEFORE finalizing your clinical evidence strategy and be sure to consider how reimbursement activities will influence your overall development timeline
- Health Technology Assessment (HTA), also referred to as value assessment, is the process of systematic evaluation of the impact, cost-effectiveness, comparative effectiveness, or cost-benefit (all synonyms), for the valuation of health technologies for purposes of decision-making
- When developing your strategy for value assessment, you need to, at a minimum, describe the following: the *disease or condition, current treatment approaches, the device or new technology, the place of the device in current treatment, the clinical effectiveness, and finally the economic benefit*
- Understanding the implications of the "coding" system for your device is imperative to plan for the reimbursement evaluation
- You need to know who is financially responsible when your device is used and who will be affected, both positively and negatively

- Generation of evidence-based data can be difficult—include the payer in the consideration of what data are needed
- Consider a provider-based, risk-sharing agreement with the payer in which a payer agrees to reimburse for use of the device while evidence is being generated
- Do not forget to plan for a more globally based market

REFERENCES

[1] Walker S, Sculpher M, Claxton K, Palmer S. Coverage with evidence development, only in research, risk sharing, or patient access scheme? A framework for coverage decisions. Value Health 2012;15(3):570–9. https://doi.org/10.1016/j.jval.2011.12.013.

[2] Neumann PJ, Chamber J. Medicare reset on 'coverage with evidence development'. In: Health Affairs Blog; April 1, 2013. Available at: http://www.healthaffairs.org/do/10.1377/hblog20130401.029345/full/. Accessed 7 October 2017.

Chapter 17

Market Adoption of Innovation Into the Operating Room: The "Hospital Chief Financial Officer as the Customer"☆

Ryan D. Egeland*, Zachary Rapp† and Frank S. David†
**Medtronic, Plymouth, MN, United States, †Pharmagellan LLC, Milton, MA, United States*

Anything that won't sell, I don't want to invent. Its sale is proof of utility, and utility is success.

Thomas Edison [1]

The single necessary and sufficient condition for a business is a paying customer.

Prof. Bill Aulet, Martin Trust Center for MIT Entrepreneurship [2]

These are mixed times for entrepreneurial physician-innovators [3,4]. Funding of private medtech companies (devices, diagnostics, and other tools and supplies) by venture capital increased by 10% in 2016, reaching $5.6 billion. In addition, past R&D investments in the medtech have been productive; there were 51 premarket approvals (PMAs) and humanitarian device exemptions (HDEs) awarded to new, innovative medical devices by the US Food and Drug Administration (FDA) in 2015. Investments in medtech companies are spawning development of an ever-increasing number of new and innovative solutions to problems in health care.

Despite this ostensibly healthy overall trend in medtech product development, entrepreneurs and industry as a whole are facing ever-increasing challenges in this very competitive marketplace. In the past, the focus of many physician innovators has been on enhancing technical capabilities by improving

---

☆. This report was adapted from a previously published article: Egeland RD, Rapp Z, David FS. From innovation to market adoption in the operating room: the "CFO as customer". Surgery 2017;162(3):477–82.

Medical Innovation. https://doi.org/10.1016/B978-0-12-814926-3.00017-6

intraoperative techniques to obtain better diagnostics and medications to achieve more meaningful, far-reaching patient outcomes [5]. Currently, however, addressing only the surgeon, the physician, and the patient's needs is only one of the criteria of value analysis committees in both hospitals and large medical institutions. The introduction of many new, innovative products is incompatible with budgetary restraints facing actual buyers when all aspects are considered—that is, the need for education of the users, replacement of current products/devices, the impact on operative time, and the multitude of other costs involved in replacing obsolete products.

Financial hospital stakeholders insist that clinical superiority is still necessary to drive the adoption in the operating room or the clinic of any new device or product. This means that no longer, however, is clinical superiority alone sufficient for the adoption of many of these innovations. Chief financial officers (CFOs) and other members of the C-suite, members of value analysis committees, departmental administrative managers of operating rooms and clinic expenses, and procurement specialists—all of whom are the ultimate "buyers" of innovative products—are struggling to manage the challenges of adopting new, attractive products into their formulary if the products add cost to the bottom line.

There is, however, a potential silver lining. Many of these innovative medical devices can and will be successful in the clinic or the operating room—provided the physician innovators expand their arguments of clinical value to include also the now very important economic value to financial stakeholders. This mindset requires an appreciation by the entrepreneur of the CFO as a prospective "customer," such that innovators first focus their innovation efforts on products that improve efficiency and efficacy while improving patient care. Using this approach, such products offer a marketing advantage, and thereby help to drive market adoption as well as serving to attract patients to the institution. Through insights from multiple interactions with hospital finance professionals, we outline a strategy for incorporating this approach into the mindset of the innovator early on when designing products for the operating room.

PRESSURES ON THE OPERATING ROOM (OR) FINANCIAL COMMITTEES ARE SERIOUS AND ACCELERATING

Despite past positive budgetary successes, most hospitals in the United States are now noticing financial constraints [6]. Due to demographic shifts and consequences of the Affordable Care Act, as well as the worries of the current changes to come in health care funding, federal programs cover a greater percentage of patients with reimbursable payouts, but reimbursement will not match the true increase in costs of care for many or most of these programs [7]. To address this, hospitals have adopted increasingly aggressive measures to promote efficiency and cut unnecessary expenditures [8]. Although financial margins (revenue minus expenses) for many hospitals have stabilized, the Congressional Budget Office warns, "(The) recent trend does not necessarily provide a good indicator of hospitals' financial health in the longer term." [9]

The hospital operating room for inpatients encompasses about 40% of hospital surgical procedures. Because of changes in how hospitals are to be reimbursed, the costs of operating room procedures may be susceptible to unique fiscal pressures [5]. Medicare reimburses a fixed amount for all inpatient operations and other procedures. Under this system of capitation, the reimbursed amount is based on estimates of both cost and complexity; however, financial experts describe the costs of operations and subsequent postoperative care—which includes labor, infrastructure, consumables, and capital equipment—to have increased much faster than the reimbursed rates. This unequal increase in cost has eroded hospital margins, making it more difficult for them to adopt innovative products with higher costs [10,11]. The following explanation came recently from a CFO at a Mid-Atlantic community hospital, "It's becoming tighter and tougher. Medicare rates are about the same as they were in 2010, so we're very judicious about what we approve. I have used the word 'no' more times in the last three years than in the previous 20 years." While private insurance companies may offer somewhat greater reimbursement rates, margin pressure is active here as well [12].

These current realities combined with concern about future reimbursement have led many hospitals and institutions to become much more attentive to the costs of inpatient and outpatient surgical technology unless there is an *objective cost savings* included in their value proposition. Unfortunately, even some of the most clinically valuable products are unlikely to be approved by some hospitals and medical systems if they add substantial costs not explicitly reimbursed. As an example of this, the director of the supply chain management at a Northeast community hospital reacted to a new device for a thoracic surgical approach to treating asthma by saying, "Reimbursement wasn't going to come close to covering the consumables," he explained. "It was clearly something that would be good for patients, but we'd take a loss on every case." With this in mind, the hospital's decision was that it could not afford to adopt the new product.

Similar and possibly realistic words for new device companies that produce new surgical products came from another CFO at a community hospital, "The device industry traditionally made money on a simple business model—come out with a replacement product that costs three to five times more than the old one. But the ability to do that has gone away. My advice is, don't spend money developing more expensive devices. We're not going to buy them."

A MYTH OF "VALUE" IN THE OR SOMETIMES CONFLICTS WITH FINANCIAL REALITY

As the costs of health care delivery have increased beyond reimbursement, many companies that manufacture drugs and devices now understand the concept of "value" as measured by the difference between the total costs involved in the adoption into the market and the "value" they create in the economics of health care delivery [13–15]. This calculus underscores the basis of the analyses of cost effectiveness employed in the United Kingdom to evaluate new therapies. Under this system, new therapies, devices, and drugs are adopted only if the monetary value of their long-term effects (often measured by what is termed "quality-adjusted life years" or QALYs) exceeds their price [16].

Value-based payments are gaining some traction in selected areas of health care delivery in the United States [17], such as orthopedic and cardiac surgery [18]. Unfortunately, most operative procedures currently are still reimbursed according to the payment schedules of the diagnosis-related groups (DRGs). Innovative devices, even if expensive, may very well not increase the DRG payment schedule. This situation creates an important and often unclear debate between "value" and "affordability" in the operating room as well as other hospital service lines. Warren Buffett has said, "Price is what you pay; value is what you get." [19] Under our current system of capitated payments for many patients, hospitals end up paying for the added costs of new, OR-based technologies, while the patients, the payers (insurance companies), and society get the benefits of better clinical outcomes and less overall costs of health care.

This scenario is especially true for new surgical products or devices. The primary concern of the hospitals is cost; long-term calculations (greater than 1 year) of health benefits from an economic basis may play little or no role in the decision-making of a hospital. Indeed, a hospital executive in the management system of the supply chain at a Mid-Atlantic academic hospital said, "We're open to looking at [health economic] analyses, but those are soft dollars. They don't have much value."

This argument of "value" should not be interpreted as suggesting that surgical device innovators and other entrepreneurs should ignore opportunities to develop, market, and deliver innovative, long-term improvements in health care. But current financial realities simply restrict the decision-making autonomy that many or most hospitals have in adopting new operating room technologies. Thus, over the near term, hospital finance executives will continue to struggle to afford many innovative products, even when these new products offer high clinical and long-term health economic value. It is therefore, imperative for innovators to make strong, evidence-based financial justification for the adoption of their products.

HOSPITALS *SEEK* TO PURCHASE SPECIFIC TYPES OF SURGICAL INNOVATIONS

Current innovators must understand the key economic customers for new devices or treatments—the CFO and others in the financial arm of the various institutions—whom they plan to target.

But who is the real "customer" of new surgical devices? Historically, most manufacturers focused their marketing efforts almost exclusively on clinicians and patients, simply because they, the end users, appeared to be the groups who benefited the most from their innovation. The decision to develop, market, and commercialize a new technology was based primarily on its ability to help physicians achieve a better-quality outcome, generally meaning better quality of life or enhanced survival for the patient.

Although surgeons and patients can provide an important and crucial perspective on new devices and treatments, they are not the sole "customers." The surgeon and the patient often bear little to none of the direct costs related to new products. Although they represent important users and the direct beneficiaries of these new medical products, their share of influence on the value analysis committee and financial executives who often determine adoption of new innovative approaches is often limited [20]. Ultimately, it is these financial managers who hold the purse strings.

Within this arena, the "CFO as customer" has become much more relevant and even dominant in this decision-making process. The days of hospital analysis committees automatically approving "surgeon preference items" are vanishing. In contrast, financial managers require clinicians to justify the added expense of new technology. One CFO of a large suburban hospital said, "If there's a major price difference, we can show the surgeons the numbers and then they sometimes say, 'The increase in quality is not worth that much.'"

The most vulnerable new products are those that are qualitatively "better," but fail to offer advantages that can be easily captured and quantified financially. As a regional hospital executive stated, "Quality of the product is almost a non-existent factor. It's purely dollars and cents—you're competing for dollars in the OR budget with minimal involvement of what the product brings to the table." Stated differently by another CFO, "The efficacy argument is hard to make. It's easier if there's an offset in terms of time reduction or something else."

This scenario leads to the importance of recognizing the "CFO as customer," in other words improving care without busting the budget. Although many innovative surgical devices claim to offer hospitals incremental revenues through a marketing advantage in attracting patients from the community [21], financial stakeholders generally demand that the increased cost of a new product be balanced by tangible, objective, near-term savings in areas that they can understand, quantitate, and manage (see Fig. 17.1).[1] Specifically, CFOs and other hospital finance professionals highlighted three ways a new surgical product could be most financially attractive to them:

1. Although a surgical device could provide other sources of value to a hospital, like revenue from new procedures or increased patient attraction or retention, finance stakeholders told us that for most new operating room products, these sorts of potential benefits are less compelling to them than near-term cost offsets.

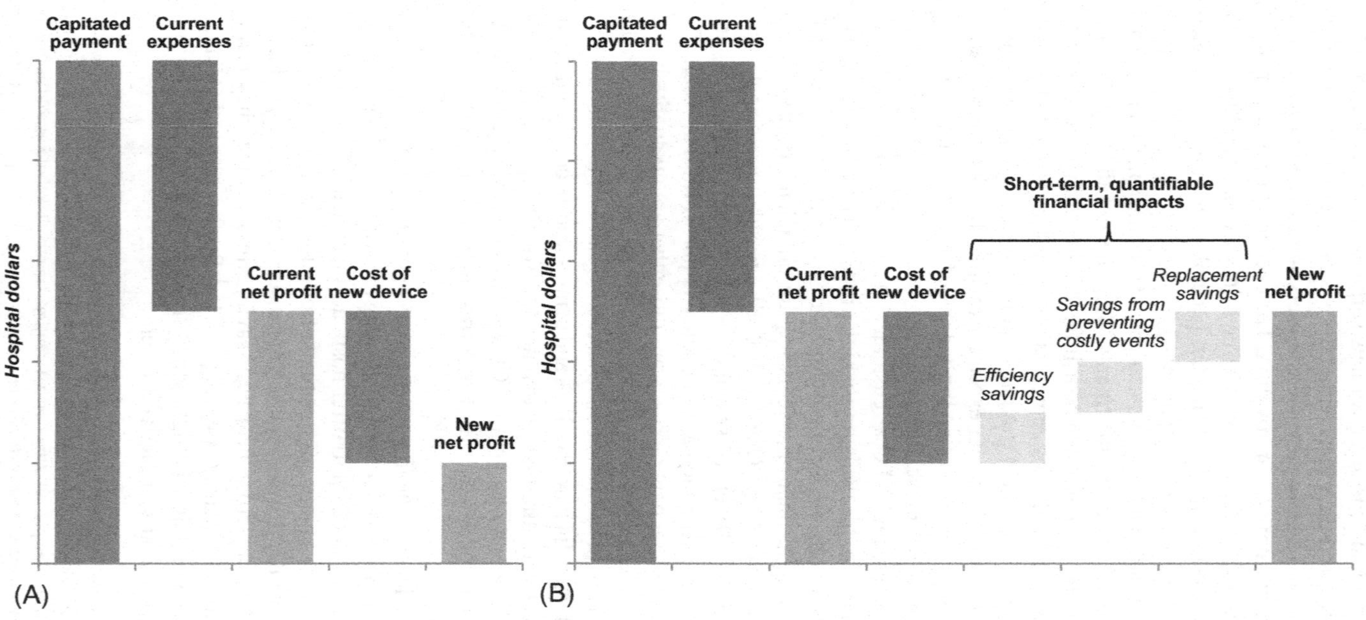

FIG. 17.1 Illustrative sources of near-term, quantifiable financial value for a surgical device. (A) Under capitated reimbursement, the hospital receives a fixed payment for a case, from which its expenses are subtracted to yield current net profit. The cost of a proposed new device is an additional expense, which decreases the net profit. (B) Cost savings due to increasing efficiency, preventing costly events, and/or saving money through substitution can restore net profit—even, in theory, to a level at which the savings reach or even exceed the cost of the new device.

BOOST EFFICIENCY

When devices can decrease *substantially* the duration of procedures or the turnover and the down time between procedures, an incremental number of procedures may be performed in the fixed hospital infrastructure. This possibility may enhance profitability in obvious ways to financial managers [22]. As stated by one manager, "If we can get the next patient in [the operating room] 30 minutes earlier, thats a big deal for us." On occasion, a realistic and convincing argument for saving time can exist beyond the times in the operating room in the postoperative period, because the added cost per day (or even per hour) typically falls under a capitated payment. One supply chain executive of a large ambulatory surgery center said, "[A device that] reduces length of stay is a no-brainer." Finally, because personnel represent often a very large percentage of costs and expense (in some situations the greatest overall category of expense), any surgical device that can decrease staff requirements will be particularly attractive to CFOs in hospital and ambulatory care centers [23].

Prevent High-Cost Events

The economic value of a surgical device that prevents undesirable complications or need for readmission is calculated as the total cost of the adverse event multiplied by the number of events expected to be prevented [24], assuming the cost of the event is borne by the purchaser. For example, the cost of a postoperative readmission within 30 days of discharge usually falls under capitated reimbursements for the initial admission and therefore, decreasing the number of readmissions will have objective, attractive, and very tangible savings (i.e., value) for the hospital.[2] This "value of prevention" can be the most compelling argument for well-researched events previously quantitated by the hospital in terms of incidence and associated costs.

Replace Expensive Goods With Cheaper Ones

Innovation that delivers identical or improved clinical capabilities at a decreased cost (or via enhanced quality and reliability) is attractive to CFO customers. For example, on occasion, new devices in the operating room may fit the model of the "LED light bulb," whereby a cheaper but less durable product is replaced with a new, more expensive one that lasts considerably longer and is considerably more efficient. Other innovative opportunities may "replace" current technology such as device usage in a completely different class of products. For instance, a type of new physiologic monitor that would

2. In contrast, decreasing readmissions (or reoperations) at further dates in the distant future, while a worthy goal with financial benefits very the long run, does not translate directly into immediate benefits financial managers seek to capture.

allow surgeons and anesthesiologists to better regulate their use of fluids, drugs, or other products can avoid wasteful over-administration in addition to yielding superior clinical benefits such as decreased complications.

A key fact that unites the three categories is that *the CFO defines value as a positive, short-term, quantifiable financial impact on the hospital.* All three of those components—short-term, quantifiable, and financial—are critical. CFOs generally value *short-term* savings (within a budget quarter, or at longest, a fiscal year) much more highly than promised longer-term benefits (greater than 1–2 years or over a lifetime), which are harder to predict or measure accurately. These savings must be *quantified* using rigorous, accepted data and metrics that reflect real-world hospital economics. Finally, although benefits to patients, physicians, and others in the health care system are important "table stakes," CFOs focus on responsibly controlling the *financial impact* on the hospital budget.

UNDERSTANDING CFOs WILL ENHANCE THE OVERALL IMPACT OF SURGEON INNOVATORS

Despite the pressures on hospitals to make ends meet, surgery and many of the medical specialties remain a fertile territory for medical innovation. Passionate and creative practitioners will continue to discover viable solutions to clinically important problems; the key is to convince colleagues and CFOs that many of the innovative new devices and technologies will both improve outcomes and patient care AND at the same time decrease the cost of care.

To have the broadest clinical impact, however, surgical device innovators need to consider, understand, and then focus on ideas that will succeed not just clinically but also financially in the hospital-based marketplace. This concept requires focus on the economic needs of customers, not only the patients and the physicians, but also the hospital financial committees and CFOs who actually "buy" these new technologies. Those technologies that deliver better clinical *and* economic outcomes will be attractive to the widest set of customers and prove much more likely to succeed in the current, financially constrained hospital markets.

Therefore, both established innovators and entrepreneurs—and even more those new to commercialization—should evolve their approaches along two lines. First, they must commit time and effort into understanding the role and outlook of the "CFO as customer"; yes, clinician champions and patients are important, but without the buy-in of the actual purchasers, the ability to move from concept to commercialization may very well fail. This concept means that the innovator and entrepreneur must "get out of the building" [25] and listen to and learn from their colleagues in finance, procurement, materials management, and the business administration in their department, hospital, or institution when it comes to understanding how the OR, the value assessment committees, and CFOs function financially. This approach will allow innovators to first understand and then to create the economic evidence the CFO needs to evaluate both the financial and clinical merits of any new device or innovation. This approach means innovators must apply the same rigor in analysis of the financial implications of their products as they apply to proving clinical benefit.

Second, the successful surgical innovator will understand what to work on, but even more importantly, what *not to work on* early in the process of concept development. Because time is valuable and limited, ideas that have little chance of being adopted into the hospital or clinic marketplace due to financial obstacles may prove to be a waste of time. Entrepreneurs will find that time devoted into researching some of the methods that successful tech entrepreneurs have used to refine and prioritize new concepts may prove very worthwhile [26].

In the operating room as well as the clinic, innovations directed toward cost-reduction and fiscal efficiency will reap greater rewards in the marketplace than those directed toward purely clinical or technical advances. Through this lens, innovators will give their technologies the greatest chance of attracting funding, partnerships with investors, and established medical device companies, and ultimately a possible chance of being adopted broadly. As Thomas Edison observed years ago, the adoption of an innovative new product by the paying customers is the surest sign of success.

IMPORTANT POINTS

- Understanding that the CFO, the C-Suite, and the value analysis committee ARE the "customers" of new technologies is imperative for the entrepreneur; this concept must be understood early in the process of medical innovation.
- Innovations need either to save the hospital/institution money or truly add to enhanced survival or quality of life, but the arguments supporting these advantages need to be focused with evidence-based data or financially justified support.
- The following three means by which new surgical products will be financially attractive
  (1) Boost efficiency
  (2) Prevent high-cost events
  (3) Replace expensive goods with cheaper ones
- To have the broadest clinical impact, innovators and the innovator/entrepreneur need to consider, understand, and then focus on ideas that will succeed not just clinically but also financially in the marketplace.

ACKNOWLEDGMENTS

We thank Bill Brosius (Baylor St. Luke's Medical Center), Michael Burke (NYU Langone Medical Center), Bill Garwood (Angleton Danbury Medical Center), David Neustaedter (Medtronic), Amy Pollack (Medtronic), and Bijan Salehizadeh (NaviMed Capital) for their helpful comments.

DISCLAIMER

No direct or indirect financial support was provided to the authors for their contribution to this report. This analysis focuses on the United States, which is the dominant source of revenues for most global medtech companies.

REFERENCES

[1] Woodside M. Thomas A. Edison: the man who lit up the world. New York: Sterling; 2007.

[2] Aulet B. Disciplined entrepreneurship. Hoboken, NJ: John Wiley & Sons; 2013.

[3] Pulse of the industry: medical technology report 2016. Ernst and Young Global Ltd. http://www.ey.com/gl/en/industries/life-sciences/ey-vital-signs-pulse-of-the-industry-2016 [accessed 13 February 2017].

[4] Cairns E, Armstrong M. Medtech 2015 year in review. Evaluate Ltd. Available at: http://info.evaluategroup.com/epv-mr-15-cs.html [accessed 10 February 2017].

[5] Riskin DJ, Longaker MT, Gertner M, et al. Innovation in surgery: a historical perspective. Ann Surg 2006;244:686–93.

[6] Egeland RD, Rapp Z, David FS. From innovation to market adoption in the operating room: the "CFO as customer". Surgery 2017;162(3):477–82.

[7] Trendwatch chartbook. American Hospital Association; 2016, http://www.aha.org/research/reports/tw/chartbook/index.shtml [accessed 13 February 2017].

[8] L.E.K. Consulting. Strategic hospital priorities study: hospitals look to medtech for new services and solutions, vol. XVI(5). L.E.K. Executive Insights; 2014, http://www.lek.com/our-publications/lek-insights/hospitals-look-to-medtech-medical-device-for-new-services-and-solutions [accessed 13 February 2017].

[9] Hayford T, Nelson L, Diorio A. Projecting hospitals' profit margins under several illustrative scenarios. Congressional Budget Office working paper 2016-04, https://www.cbo.gov/publication/51919; 2016 [accessed 13 February 2017].

[10] Surgeons and bundled payment models: a primer for understanding alternative physician payment approaches. American College of Surgeons; 2013, https://www.facs.org/~/media/files/advocacy/pubs/bundled%20payments.ashx [accessed 13 February 2017].

[11] Adopting technological innovation in hospitals: who pays and who benefits? American Hospital Association; 2006, http://www.aha.org/research/policy/2006.shtml [accessed 13 February 2017].

[12] Reinhardt UE. The pricing of U.S. hospital services: chaos behind a veil of secrecy. Health Aff 2006;25:57–69.

[13] Porter ME, Kaplan RS. How to pay for health care. Harvard Bus Rev 2016;94:88–98.

[14] Value-based healthcare: Strategies for medtech. The Economist Intelligence Unit; 2014, www.eiu.com/vbhmedtech [accessed 13 February 2017].

[15] Taylor M. The role of health economics in the evaluation of surgery and operative technologies. Surgery 2017;161:300–4.

[16] Appleby J, Devlin N, Parkin D. NICE's cost effectiveness threshold. Br Med J 2007;335:358–9.

[17] CMS's value-based programs. Center for Medicare and Medicaid Services. https://www.cms.gov/Medicare/Quality-Initiatives-Patient-Assessment-Instruments/Value-Based-Programs/Value-Based-Programs.html [accessed 13 February 2017].

[18] Bundled payments for care improvement (BPCI) initiative: general information. Centers for Medicare and Medicaid Services. https://innovation.cms.gov/initiatives/bundled-payments/ [accessed 13 February 2017].

[19] 2008 annual report. Berkshire Hathaway, Inc. http://www.berkshirehathaway.com/reports.html [accessed 18 December 2016].

[20] Kotler PT, Armstrong G. Principles of marketing. 16th ed. Essex, UK: Pearson Education Ltd.; 2016.

[21] Shortell SM, Morrison EM, Hughes SL, et al. The effects of hospital ownership on nontraditional services. Health Aff 1986;5:97–111.

[22] Macario A. What does one minute of operating room time cost? J Clin Anesth 2010; 22(4):233–6.

[23] Kaplan RS, Haas DA. How not to cut health care costs. Harvard Bus Rev 2014, https://hbr.org/2014/11/how-not-to-cut-health-care-costs [accessed 13 February 2017].

[24] David FS. Calculating the value of an EpiPen. In: Pharmagellan blog. 2016, http://www.pharmagellan.com/blog/epipen [accessed 13 February 2017].

[25] The lean approach. Ewing Marion Kauffman Foundation. https://www.entrepreneurship.org/learning-paths/the-lean-approach [accessed 13 February 2017].

[26] Chen E. Think like a founder before becoming one. Xconomy; 2014, http://www.xconomy.com/boston/2014/03/28/think-like-a-founder-before-becoming-one/ [accessed 13 February 2017].

Chapter 18

Accelerating Physician Entrepreneurship: Perspective of a Recently Graduated Medical Student

C. Corbin Frye
General Surgery Resident, Washington University School of Medicine, St. Louis, MO, United States

Coming into medical school I was pretty naïve. I assumed that one went to medical school to become *just* a doctor. However, medical students are surprised each year to encounter diverse experiences that open their eyes to the variety of different options and career paths available to them as soon-to-be physicians. In the classroom, students realize the impact physician-educators have by teaching the next generation of doctors. Many trainees study disease states and publish research with physician-scientists. Through the American Medical Association, students recognize the importance of having physicians involved in the legislative process. Trainees work with physicians that hold roles in hospital administration. Personally, I went on a surgical mission trip to Haiti which confirmed my own desire to be a surgical missionary. I also sat on my school's curriculum council and collaborated with attending physicians to improve academic policy and educational curricula. Even, however, with all of these options on the table, there was still a feeling that something was missing. I was drawn to medicine because I loved the personal interaction and joy that came with taking care of people and, while I could not categorize it at the time, I had an inquisitive and entrepreneurial mindset that enjoyed the thrill and satisfaction of bringing a concept to fruition. But I had no idea how to bridge the gap between the two.

Then I discovered medical innovation. I was first exposed to the concepts of health care entrepreneurship and biodesign during a residency interview in which I had the opportunity to hear from and speak to physician-innovators who made their careers in both taking care of patients *and* in furthering medicine by designing solutions to relevant problems in health care. Many medical students are led to believe that there is a clear and immobile boundary between the business and sales tactics of "industry" and the art of healing practiced by

Medical Innovation. https://doi.org/10.1016/B978-0-12-814926-3.00018-8

physicians. In short, most medical-trainees do not even know that being a physician-innovator is an option. Contrary to what the broader medical culture had taught, I realized that as a physician-innovator I could use my driven and always-questioning personality to think critically, challenge myself to create something new, and have a positive impact on the way that medicine is practiced.

It is clear to me that there is room for improvement in exposing and nurturing medical students to become involved in medical innovation and entrepreneurship. While the issue is multifaceted, I would like to propose the following question of which the first part of this chapter will be devoted to discussing, exploring, and with any luck, helping to answer: "What changes need to occur in order to create an environment in which health care innovation and entrepreneurship can thrive within the greater context of medical education?". There are two arenas in which we, as a medical and innovation community, have opportunities to create change in an attempt to bring innovation to the forefront of medical education:

Part One: Advancing Health Care Innovation in the Context of Medical Education

1. Opportunities for Improvement in the Mentality and Culture of Medicine
2. Opportunities for Improvement in the Medical School Experience

In the second part of the chapter, I hope to provide medical trainees at all levels with some general advice by reviewing some of the pearls, pitfalls, and lessons that I learned during my own journey of becoming a medical trainee-entrepreneur. While the advice and discussion in this chapter is aimed specifically at medical students, many of these topics and lessons can be applied more broadly to any would-be entrepreneurs.

Part Two: Thoughts and Advice for Medical Trainee-Innovators

3. Reflections on Medical Student Innovation
4. Entrepreneurial Lessons I Learned Along the Way

PART ONE: ADVANCING HEALTH CARE INNOVATION IN THE CONTEXT OF MEDICAL EDUCATION

Opportunities for Improvement in the Mentality and Culture of Medicine

Business management educator and author, Peter Drucker, famously said that "culture eats strategy for breakfast." If biodesign and medical innovation are going to play a more substantial role in medical education, the culture of the medical community must change such that we realize that *every* medical student can benefit from learning about medical innovation and entrepreneurship. While the majority of medical students will not go on to start companies, design medical devices, or research health services technology, medical students will be better physicians for having studied these topics. Of the six core competencies of the Accreditation Council for Graduate Medical Education (ACGME), four of them—Practice-Based Learning and Improvement, Systems-Based Practice, Professionalism, and Interpersonal Skills and Communication—can be developed directly in medical trainees through learning opportunities revolving around medical innovation. Those who study business management and leadership will likely be better employees, understanding more completely—and thus being able to contribute more effectively to—the organizational goals of their hospital or group practice. From a practical standpoint, education on finance and business fundamentals will teach students about the complexities of health care economics, while also preparing a subset of students that will one day own a medical practice or develop a career in hospital administration. Through the study of entrepreneurship, doctors-in-training will learn how to better differentiate themselves in the increasingly competitive health care market and will learn how to *create* a job for themselves rather than find one. Students will gain a greater appreciation for the intricacies of medicine by learning how health care devices are developed and how medical services are delivered. Through innovation studies, students will learn how to think outside of the box, question the status quo, and communicate effectively. Innovation teaches problem solving, creativity, collaboration, leadership, grit, and determination. Students will learn how to identify problems, verify assumptions, and keep an open mind about how they can improve patient care and satisfaction. Put plainly, students who study innovation, entrepreneurship, and biodesign will be better doctors for it.

In modern-day medical culture, there is a commonly held perception that the healing art of medicine does not and cannot mix with the business and financial foundations of entrepreneurship. There are many examples, however, of medical innovation changing the landscape of health care and improving the lives of patients all across the world. Cook Inc. was founded out of a spare bedroom by Bill Cook in my hometown of Bloomington, Indiana, in 1963 with an investment of just $1500 [1]. Mr. Cook was a curious and passionate person who had no idea that his work with wires and catheters would help spawn a whole new field of minimally invasive procedures that would allow for the treatment of many patients that otherwise faced morbid and often unsuccessful interventions. While “industry” has the stereotype—which is in some instances well deserved—of being focused solely on the bottom line, that certainly is not a requirement of those who participate in the field of health care innovation and biodesign. I firmly believe that the altruistic values of medicine can align with medical entrepreneurship, and in fact, this is a large part of my personal ambition of pursuing medical innovation. As a surgeon, I will be able to take care of a handful of patients each day, but as a physician-innovator, I have the potential to have a positive impact on hundreds or even thousands of patients every day by developing a medical device or creating a business that solves a crucial health care enigma. You need not aspire to run a Fortune 500 company to have an interest in medical innovation. Like Mr. Cook, you just have to have a heart for improving patient care, a curious passion for solving problems, and an unwavering dedication to keep you going when times get tough.

Opportunities for Improvement in the Medical School Experience

Medical schools not only have a role in providing trainees with their primary medical education, but more broadly in creating an environment in which students can discover different career options, explore new disciplines, and grow and develop as future physicians. Not surprisingly, there are certainly opportunities within the medical school experience to improve trainees' knowledge of and interaction with innovation, medical entrepreneurship, and biodesign.

The most straightforward way to achieve this is for medical schools to broaden the scope of education they provide by building innovation and biodesign directly into the formal medical school curricula. While there certainly are complex and vying demands on medical school curricula, one way to create educational opportunities without depleting too many resources would be offering a seminar-style course titled "Foundations of Medical Innovation, Biodesign, and Entrepreneurship" for all medical students that is pass-fail and only occupies a few credit hours. Medical schools could bring in physician-innovators and business school professors, along with local experts, such as lawyers, engineers, and entrepreneurs, to give discussion-based lectures on a variety of topics, including bioengineering, medical-legal studies, business and financial foundations, entrepreneurship, and leadership. This class would not only generate increased interest in the subset of would-be student-innovators, but more importantly for the medical student class at large, this approach would create a more informed body of soon-to-be doctors who are better able to appreciate the complexities of health care economics and the development of medical devices and health care services.

Another educational opportunity would be the creation of fellowships in innovation and biodesign in which medical students could gain focused training, mentorship, and most importantly, structured, hands-on experience in the fields of health technology innovation and medical entrepreneurship. These types of fellowships already exist at the graduate and professional levels. Examples include the Texas Medical Center Biodesign Fellowship [2] and the Pediatric Innovation Fellowship at the University of Michigan [3], but the "first-on-the-scene" and well-accepted gold standard has been the Stanford Innovation Biodesign Fellowship [4]. In the Stanford fellowship, participants walk through the entire innovation cycle in a 1- or 2-year immersive experience that includes need identification, concept development, assumption validation, business planning, and even fundraising and company building. At the medical school level, these fellowships could be made more attractive by providing opportunities for academic productivity and publishing in the literature or even including a personal stipend potentially funded through state and federal agencies, such as the NIH. Currently, it is not uncommon to see medical students devote 1–4 additional years during medical school to focus on academic productivity in a laboratory or to pursue other degrees, such as a PhD, MPH, or MBA. To that end, these potential fellowships, similar to the one at Stanford, could require a year or two of additional devoted education. Alternatively, medical schools could run a smaller scale "mini-fellowship" in parallel with existing curricula in which participants could avoid delaying their anticipated graduation date. Regardless of the details, providing these types of formal training opportunities in health care innovation and entrepreneurship during the early years of medical education would serve to jumpstart the careers of interested student doctors.

Medical schools also have the opportunity to inspire innovation through relevant extracurricular endeavors outside of the classroom, including events such as multidisciplinary workshops, in which a university could pool leaders from the schools of medicine, law, business, and engineering to provide TED talk-style seminars on related topics in innovation. These same events could also serve as "meet and greet" sessions, in which the leaders and students from each of the disciplines network and collaborate on ideas or projects of their own. Another avenue is through supporting student-led medical entrepreneurship clubs or "Student Interest Groups," in which students could host lunch talks, networking events, skills workshops, or lectures on topics pertaining to entrepreneurship and biodesign in health care.

The best way, however, for medical trainees to learn medical innovation is not through reading a textbook or taking a course but through first-hand experience. Medical schools could encourage and facilitate these types of real-world experiences by providing resources and opportunities for medical students who are interested in solving a particular relevant problem in health care or perhaps are already in the early stages of their own startup company. One such resource is the entrepreneurship incubator. Through partnership with the medical school's Office of Technology Management or Office of Technology Transfer, these university-run incubators provide participants with direct mentorship, venture and market analysis, business development, legal and patent resources, networking, and education about intellectual property. Frequently, these types of incubators offer opportunities for funding through business plan competitions, grants, or venture capital. One example of an established university-associated entrepreneurial incubator is the LEAP (Leadership in Entrepreneurial Acceleration Program) Inventor Challenge at Washington University in St. Louis [5]. In this program, teams work on a particular translational idea and receive personal mentorship, guidance, and analysis from experts at the Skandalaris Center for Interdisciplinary Innovation and Entrepreneurship [6]. Specifically, teams are able to develop their concepts and businesses by meeting with venture analysts to probe for potential weaknesses and opportunities in their business plans, creating a fully vetted and revised executive summary, and even getting visual design help on making an esthetic, streamlined "pitch deck." The program ends with a competition in which teams pitch their ideas to judges, with the winning teams receiving funding to further develop their ideas. Win or lose, those who participate in these types of programs walk away with new mentors and colleagues, a more complete understanding of their particular idea of interest, and most importantly, with the first-hand experience of partaking in the process of innovation and entrepreneurship.

Finally, there is room for medical schools to help make medical innovation more of an academic pursuit. Some find it hard to believe that entrepreneurship and health care innovation can be successful at an academic institution, but in fact these endeavors can perhaps *best* be explored in an academic setting, where medical research and the improvement of patient care are foundational to the institutional mission. Thus, medical schools, with their vast array of financial and multidisciplinary resources, are places where medical innovation can naturally flourish. An important and often highly sought-after experience for many medical students is the participation in academic research labs and the subsequent productivity of publishing in the academic literature. Medical schools have the opportunity to encourage and support the addition of academic "bench-to-bedside" translational research labs that focus primarily on biodesign and innovation. One of my own mentors at Washington University, vascular surgeon Dr. Mohamed Zayed, successfully created two companies during his residency training and is now the principal investigator of his own research lab focusing on the development of new therapeutic modalities for the treatment of peripheral arterial disease in diabetic patients [7]. Like other academically focused physicians, Dr. Zayed applies for grants and other funding, but with his laboratory's resources, he helps lead medical trainees through the process of need identification, solution designing, and if successful, translation to market. Not only are future patients benefiting from Dr. Zayed's work, but the participating medical trainees are experiencing first-hand the many steps involved in taking an innovative idea in health care from concept to commercialization, while still being productive from an academic and research standpoint. Translational labs that research and create new health care devices, technologies, and services will forge a more innovative environment, while also generating further interest in academic medicine, specifically from trainees who are less interested in traditional "bench top" or basic science research.

PART TWO: THOUGHTS AND ADVICE FOR MEDICAL TRAINEE-INNOVATORS

Reflections on Medical Student Innovation

One area in particular that medical students interested in medical innovation need to have more intentionality is in seeking entrepreneurial mentorship. Someone once gave me the advice: "treat mentors like baseball cards: collect as many as you can and hold on tightly to them." Mentorship only rarely occurs on its own, so you have to be deliberate about seeking out advisors, but in my experience, the fruit of mentorship is abundant. Strong mentorship almost always has a positive, if not *career-influencing*, effect on students. Securing mentorship can be quite intimidating, but I have found that the best approach is to simply reach out to the potential advisor and ask for their guidance in plain language. Many are apprehensive about asking for mentorship, because of fear of being burdensome or suggesting a "taking advantage of" mentality. In reality, when done properly, the majority of mentors are not only happy to engage but are actually *honored* that you would think so highly of them as to ask for their guidance in your life. Specifically, trainees should consider seeking guidance from doctors at their medical center who have a startup of their own, local businessmen and women, and academic professors whose translational research lab focuses on biodesign. As a medical trainee-entrepreneur, these mentors will be critical to you by providing feedback on your ideas and by helping guide you through the arduous and often frustrating process of taking an idea to market.

Counter to popular belief, one does not have to have to have an MBA, a PhD in engineering, or a JD to start a biomedical or health care technology company. Interested medical students, however, are behooved to pursue a fundamental set of skills and understanding about innovation in addition to their traditional medical coursework. One resource in particular that has been instrumental for my own education and entrepreneurial growth is iTunes U, which provides free access to thousands of online recorded lectures and coursework from well-regarded sources. Through courses such as MIT's "New Enterprises," the University of Oxford's "Building a Business," Duke's "Bioengineering Applications to Address Global Health Conference," and Stanford's "Entrepreneurship Through the Lens of Venture Capital," one can have personal access to some of world's greatest business educators teaching the fundamentals of entrepreneurship and innovation without paying for any business classes or taking scheduled time away from medical training. There are also in-person and online nonuniversity-associated courses that one can take to learn entrepreneurial fundamentals, such as St. Louis's BioSTL "Focus on the Founders" course [8] or Ontario, Canada's MaRS "Entrepreneurship 101" [9].

Some cynics and naysayers might argue that medical trainees are too busy, too focused, or not knowledgeable enough to be productive in the fields of medical entrepreneurship and biodesign. The greatest challenge to medical students is indeed that their medical knowledge and experience in health care delivery is understandably less complete and matured compared to fully trained, attending physicians, which on the surface, may be a cause for concern. I believe, however, that medical students are *uniquely* positioned to be the future of medical innovation, because they are freshly passionate and are not bogged down with the dogma and pessimism that have been known to plague a long career in medicine. Medical students are professional learners and thinkers, and with their enthusiastic, albeit underdeveloped, vision of health care, they are distinctively poised to challenge the status quo by thinking outside of the box to solve some of the problems that modern health care faces.

Medical students should know that health care entrepreneurship can be incorporated into any type of medical career, regardless of one's specialty or professional goals. Some will exercise the option to stop clinical practice all together to pursue innovation full-time through academic research, private company building, or working in industry. The majority of physician-entrepreneurs, however, will strike a balance between their clinical practice and their time spent on biomedical innovation. Two of my innovation mentors in particular have found vastly different ways to practice medicine while still pursuing their interests in effective and efficient ways. Dr. Zayed has formally built innovation into his academic practice by leading his own research lab that explores new techniques and devices that help patients with vascular disease. In contrast, another mentor of mine, an academic Emergency Medicine physician, has used his medical knowledge to pursue entrepreneurship and innovation more informally on the side of his clinical duties, even starting his own company that is working to innovate the way that physicians clean wounds and cuts. The scope and flexibility of the health care entrepreneurship field allows the potential to tailor your involvement in it—whether you are interested in bioengineering-focused device creation, legal advising for a health care startup, or even the creation of inexpensive and durable solutions for resource-scarce areas of the world, it is reassuring to know that you can build an individualized medical innovation career around your own personal and professional ambitions.

Entrepreneurial Lessons I Learned Along the Way

My own journey as a medical student-entrepreneur was an exciting, thought-provoking, informative, and by some metrics, unsuccessful, experience. To end this chapter, I will share some of my own adventure of pursuing health care innovation during medical school and in doing so, try to impart a few lessons that I had to learn the hard way. Through discussing many of my mistakes, missteps, and occasional wins, it is my goal to provide medical trainees better insight into the innovation process. Most of all, I hope that my story can be of some encouragement to budding medical innovators trying to find their way.

A lesson that all trainee-entrepreneurs must learn early is that innovation is hard and success comes slowly. Medical entrepreneurship requires substantial time and patience. It is easy and even second nature for many medical students to want to jump right in and create a company at the first hint of a good idea. When I first heard about innovation, I hit the ground running and quickly recruited my best friend to become my business partner. We discussed a few ideas on the phone and picked one to run with. We wanted to get an LLC and a patent lawyer right away. It was later that month when a mentor helped me recognize that to do this the right way, there was a process and that it would not necessarily move fast.

I soon discovered the methodology of Customer Development—created by Steve Blank, a well-known innovation author and a serial entrepreneur himself—and it proved to be helpful in organizing my thoughts. Mr. Blank says that the four sequential steps required to have a successful company are customer discovery, customer validation, customer creation, and company building. Medical students and new health care entrepreneurs, such as myself, should focus initially on just the first two.

Customer discovery revolves around problem identification. It starts by asking if the problem you are trying to solve *actually exists*. By applying this principle, I realized the ironic fact that the first step in creating a startup business is determining if you have a business at all. Customer discovery also requires that you validate the assumptions you are making about that problem. After continuing to watch lectures and talk to mentors, I learned that assumptions equal risk and that the more risk you enter into a startup with, the more likely you are to fail. This epiphany helped my business partner and I to realize we had gotten way ahead of ourselves. We took what seemed to us to be an obvious problem in the hospital—the fact that the suction containers for nasogastric tubes had to be switched manually—and made several assumptions about the severity of the problem, who the problem affected, and most basic of all, that the problem even existed. After taking the time to talk with just a few nurses and OR technologists, we realized that what we saw as a substantial problem was really just a minor annoyance that already had an easy and free workaround.

Medical student-innovators should know that problem identification is not a step to be skipped. Once my business partner and I realized this fact, we made it our personal goal to identify as many problems as possible, spending two entire months doing nothing but coming up with and researching needs in the field of medicine. Many of the initial ideas and problems we identified came from our own personal experiences in the hospital during our clinical rotations. But many of the needs with the biggest potential for real innovative change, we identified, not by thinking out loud on a conference call, but by talking face-to-face with people on the ground. Whether it was the nurse on the inpatient floor, the janitor cleaning the operating rooms in between cases, the perfusionist running the heart bypass machine during cardiac surgery, the courier who transports blood samples, or the attending doctor, we would talk to anyone who would give us the time of day. We wanted to learn how their devices worked, what role they played in patient care, and why they did their job in a certain way. Possibly the most important question that we would ask was "What frustrates you the most on a daily basis?". From this question, we were able to complete Mr. Blank's first step, customer discovery, by learning what the real issues were to the people who would be the end users of our innovation.

The second step of Mr. Blank's pathway is customer validation, which includes asking if the proposed customer will *actually use and spend money on* the product or service of interest. My business partner and I finally landed on a problem that we thought offered a big opportunity: the surprising lack of osteoporosis screening, diagnostic, and therapeutic services available to the nursing home and assisted living populations. By working with a mentor, Dr. Jim Pike, a geriatrician-entrepreneur, we came up with what we thought was a pretty good solution to the problem. We met with yet another physician-entrepreneur mentor and immediately began pitching our idea to him and started asking about the next steps for patents and funding. He stopped us in our tracks; again, we were getting ahead of ourselves. He helped us to realize that while we had successfully identified a problem and even come up with a solution, we still had to validate several more assumptions and dig deeper to find out if we truly had something innovative that would lead to success in the real-world. This particular mentor introduced us to what I plainly call "the interview process," in which the innovator has short but highly focused conversations with the different key players in the field of interest, asking about how that person understands a particular problem and, in a neutral and nonrevealing way, asking if they think your proposed innovation model would solve the problem and be useful to the end user. Over the next several weeks, my business partner and I interviewed dozens of geriatricians, internists, endocrinologists, and nursing home administrators and executives in an attempt to validate our solution and value proposition. Through the interview process, our model received mixed but mostly positive reviews. Learning from these key players who were intimately involved in our field of interest not only helped us more fully comprehend the root of the problem itself but also illuminated several weaknesses in our model that we had not previously appreciated.

Medical training does not always encourage the same skillsets and perspectives that are required to be successful in the field of health care innovation. When it comes to the topic of failure, medical culture impresses on its trainees that defeat is not an option—missed diagnoses, complications, and bad outcomes are unacceptable. To have success in entrepreneurship and innovation, however, you not only have to accept that failure will come at some point, but you need to embrace it. Not only can mistakes make you a better entrepreneur, but they can even lead you to better ideas. This happens through iterative innovation—the notion that ideas are rarely actualized in the same form as they were conceived, but instead, require repeated cycles of trial and error before coming to fruition. For a startup company, often the first business plan or medical device or health care delivery service looks vastly different than what ends up being commercialized.

Several months of developing and hours of conversations into building our company, it was becoming clear that our business model was not going to be financially sustainable. We had pitched our idea to a couple of serial physician-entrepreneurs, crunched the numbers, read the research papers, extrapolated the diagnostics, and eventually realized that our idea was based on aging technology and that it would not be economically viable in today's health insurance markets. There was one phone call in particular in which my business partner and I were discussing the dim and uncertain future of our company. During that conversation, I recalled a Stanford business lecture given by Bill Coleman, a well-known serial entrepreneur, who was discussing the selection criteria for the next chief executive of one of his companies. I was shocked when he listed the last of the requirements: he wanted the next chief executive of his company to have previously *failed* in a startup.

The conversation with my business partner turned somber when we made the agonizing decision to officially walk away from our venture. Sighing, I said with complete sincerity, "Well, at least we can check off that requirement". We both laughed and within a few days, we were back to the drawing board discussing our next big idea. As Mr. Coleman himself said, "it is in failure that you learn infinitely more than in anything else." As in medicine, mistakes are part of the territory when you are challenging yourself to learn new skills and solve difficult problems. Medical entrepreneurship and innovation attracts dreamers who are devoted to changing the world, no matter how many obstacles or cynics they may encounter. Success in medical innovation does not necessarily require the best business skills, communication styles, or even the brightest ideas. Instead, achievement in health care innovation requires someone with a burning passion to find, and then actually solve, a problem that he or she sees in the world. There will certainly be setbacks during the innovation process, but a successful entrepreneur will be able to look at obstacles as opportunities for growth.

I once heard that "health care is a business, medicine is a science, and healing is an art." While medical innovation will not be able to resolve all the medical problems in the world, when pursued by passionate and curious people, it can make the art of healing much more joyful and impactful. At the end of the day, true innovators are not in the game for the money or the prestige, but rather for the opportunity to challenge themselves to creatively solve a problem using their passion for making a real impact on the lives around them. While my first startup did not end with any particular commercial accomplishments or accolades, I look at the experience not as failure but as feedback. For it, I am certainly a better entrepreneur and, I suspect, a better doctor too.

IMPORTANT POINTS

- Medical students will likely be better physicians if they are exposed to the principles of biodesign and medical innovation, because they will further develop their own skills in leadership, persistence, collaboration, and problem solving.
- Medical schools can do a better job of building health care innovation into medical education through seminar courses, student-led groups, networking events, innovation fellowships, and school-sponsored entrepreneurship accelerators.
- With the help of mentors and relevant resources, the best way for medical trainees to learn medical innovation is through first-hand experience.

- With the right skills and determination, building a multifaceted career that includes both innovation and clinical medicine is certainly achievable.
- Medical students who pursue health care innovation have the potential to make a real impact on the future care of patients.

REFERENCES

[1] http://pubs.rsna.org/doi/10.1148/radiol.15141243.

[2] http://www.tmc.edu/innovation/innovation-programs/biodesign/.

[3] https://medicine.umich.edu/dept/surgery/surgical-specialties/pediatric-surgery/education/pediatric-innovation-fellowship.

[4] http://biodesign.stanford.edu/programs/fellowships/innovation-fellowships.html.

[5] https://skandalaris.wustl.edu/funding/leap/.

[6] https://skandalaris.wustl.edu.

[7] http://www.zayedlab.wustl.edu.

[8] http://www.biostl.org/about/fundamentals/.

[9] https://www.marsdd.com/entrepreneurship-101-online/.

Chapter 19

Accelerating Physician Entrepreneurship: The Perspective of a Resident Entrepreneur

Kyle Miller*,†
*Bold Diagnostics, LLC, Chicago, IL, United States, †Intuitive Surgical Inc., Sunnyvale, CA, United States

Physicians are exposed to an incredibly unique professional training experience during residency that will prepare them ultimately for a career in their respective field. Residency training in some places can provide those physicians with an entrepreneurial mindset with ample opportunity to recognize unmet clinical needs in the health care arena for solutions in medical devices, digital health, and information technology (IT). Many ideas are generated, yet few physicians in training are provided with a real opportunity to create solutions addressing needs in health care that will eventually reach commercialization. Physicians in residency programs must balance a new, fast-paced work environment that results in increasing autonomy with each year of advancement. In an era of striving for a balance between workload and lifestyle, few hours remain for residents to design and create a sustainable business much less advance their clinical knowledge with each new rotation while trying to find time for family and chores. For those with drive and the passion to innovate, however, there are multiple avenues to advance their ideas, especially when made aware of the possibility for making time for professional development in programs and fellowships in innovation. This chapter outlines opportunities that are available specifically for the entrepreneur in residency training programs to help to offer methods to construct a business plan to successfully in commercialize their ideas.

RECOGNITION OF UNMET CLINICAL NEEDS

During residency, physicians throughout the country are exposed to new fields with each clinical rotation in the different specialties. Residents interested in

Medical Innovation. https://doi.org/10.1016/B978-0-12-814926-3.00019-X

innovation and entrepreneurial activities (what I will from now on call "resident entrepreneurs") can benefit from these new rotations by paying careful attention on the clinical rounds in the hospital to opportunities to improve or invent a device or other opportunities to improve health care. Current residents have grown up in the electronic era and are facile with the internet and social media and, thereby, are in a unique position to better understand and evaluate the electronic medical record, health care websites, and smartphone applications for their shortcomings, inefficiencies, and hindrances to the workflow. Further, entrepreneurial residents will also observe "workarounds" which are common in interventional suites and the operating room as well as in the clinic for novel devices and procedures. The author is a proponent of recording these unmet needs and organizing them by their respective fields (i.e., cardiovascular, gastroenterology, orthopedics, plastic surgery, gynecology, etc.) in a journal or electronic format. With each need, the entrepreneur should evaluate the population impacted and the problems encountered and consider potential solutions [1].

EVALUATING OPPORTUNITIES AND THE POTENTIAL IMPACT ON WORKFLOW

The clinical need is the ultimate backbone of the business idea and ultimately informing market opportunity and integration of the business model. In order to prioritize clinical needs for further exploration, the author suggests evaluating needs in three domains: market size, technical feasibility, and value proposition, all of which are discussed elsewhere as well in this book

For *market size*, the resident entrepreneur should investigate the existing literature to better understand the population impacted and the disease states addressed. For benign, acute, chronic, and malignant diseases, the resident entrepreneur should describe accurately the incidence and prevalence of a disease. A number of resources for investigating disease states include the administrative database of the National Inpatient Sample (NIS) (https://www.hcup-us.ahrq.gov/nisoverview.jsp), Surveillance, Epidemiology, and End Results (SEER) Program database (https://seer.cancer.gov/data/), and various meta-analyses or Cochran reviews for prevalence and incidence; market reports are available through institutional libraries that can help to define international prevalence.

Technical feasibility may require the help of an engineering colleague or professor depending on the background of the resident entrepreneur. For instance, technically challenging needs may require complex engineering solutions or scientific discoveries which should be recognized early in the need evaluation process. Engineering departments at neighboring institutions, which comprise biomedical, mechanical, electrical, industrial design, software development, and data scientists involved in the latest IT, are an incredible resource for need evaluation from a technical standpoint and engineering requirements to create potential solutions.

Value proposition is crucial and needs to be evaluated from the perspective of the various stakeholders ultimately impacted directly by the clinical need; this process may include physicians, nurses, providers, patients, third-party payers, procurement groups, office managers, device distributors, IT departments, and especially hospital administrators and their value analysis committees to better define the ultimate users and customers to be approached in the development process of the business plan. Once the evaluation criteria are established and weights for each domain are assigned by the entrepreneur, the most promising unmet clinical needs filter to the top of the list. Resident entrepreneurs are encouraged to seek input from their colleagues to confirm their assumptions prior to moving forward with the most promising unmet need.

ACCELERATING IDEATION AND DEVELOPMENT OF PROTOTYPES WITH A CROSS-FUNCTIONAL TEAM

Clearly defining the unmet clinical need allows the resident entrepreneur to enter the ideation phase where potential solutions can be explored for commercialization. This stage is best addressed with the formation of a cross-functional team consisting of members who are able to aid company formation, create and protect intellectual property (IP), and engineer device prototypes and/or software iterations and wire framing, to generate a viable, realistic business model, further explore the potential market(s), and consider regulatory and commercialization planning. Although experienced, cross-functional team members can be beneficial, there is also incredible opportunity for forming companies and generating innovative solutions around unmet needs with neighboring business, law, and engineering schools at some/many of the institutions/universities where residents are trained. Brainstorming sessions for generation of ideas for potential solutions are often most successful when unbiased members of the group are intimately involved in such sessions. Open rooms with simple whiteboards can be utilized to list as many potential solutions as possible (i.e., mechanical, electrical, software-based, etc.). Ideas can then be discussed and filtered for the need with various domain inputs, including technical implementation, IP, workflow integration, potential embodiment (for devices), enablement of efficiency for users, ease of use, and other criteria defined by the team.

IP should *not be ignored* by the group but rather embraced. IP is an important foundation for most start-up business models. Becoming familiar with existing technologic solutions for unmet needs at this point in the process allows teams to navigate the IP more effectively to generate a landscape regarding independent and dependent claims, identify companies who have carved out IP, and finally to overlay the suggested solutions with the open IP "whitespace." Google patent search (https://patents.google.com/) is a great tool to identify existing IP and should be queried similar to PubMed regarding themes and keywords. PatSnap (http://www.patsnap.com/) is subscription-based software that can provide the team with a whitespace generator, thereby saving time and potentially discovering overlooked IP that could become problematic with IP disclosures and patent filings during freedom to operate (FTO) analysis.

Rapid prototyping is an essential skill for the cross-functional team to generate "rough draft solutions" that can be evaluated by users for feedback early in the process of solution generation. Resident entrepreneurs are often at a disadvantage at this stage, hence the importance of recruiting engineering teammates with a "fail early and fail often" mindset, something that no physician without experience in innovation fully understands or embraces. The typical perfectionist personas and medical school pressure for high STEP scores in most residents hinder an entrepreneurial mindset at this stage. Further, presentations at morbidity and mortality (M&M) conferences and Monday morning quarterbacking on clinical rounds or in the operating room leave an impression that failure must be broken down and analyzed extensively. In contrast, with innovation, failure in rapid prototyping will be understood and acknowledged as a definite possibility by the experienced team allowing the group to quickly pivot toward more promising solutions. Engineering schools and local maker spaces that provide rapid prototyping tools, such as three-dimensional (3D) printers, industrial design software, tooling, and benchtop testing are a perfect environment to foster the group's creativity.

INVENTION DISCLOSURES AND LICENSING IP

Inventions by cross-functional teams involving a resident entrepreneur are often subject to the invention disclosure policy of their university or institution. Because a resident entrepreneur is, technically, supported through resources and funding within their respective institution or academic center, their inventions need to be disclosed appropriately to the technology transfer office. Although this process remains somewhat controversial with many resident entrepreneurs longing to take sole possession of ideas they generate, the university and institutional technology transfer offices can benefit the entrepreneur in multiple ways. First, inventions disclosed to technology transfer offices are often vetted thoroughly, decreasing demands on the entrepreneur's time. Further, a landscape is often generated by the technology transfer office and shared with the entrepreneur which aids the development of a competitive business plan. Technology transfer offices recognize that resident entrepreneurs may have limited resources and are thus willing to negotiate generous terms for newly formed businesses, so that a license can be obtained for rights to the IP; the cross-functional team will also evaluate the offer by the institution to determine if it is fair for the entrepreneur. Unfortunately, this will be an important consideration and requires careful attention by an experienced member of the team. Obtaining an exclusive license for the rights to IP can be a tedious process requiring company formation and achievement of milestones in order to maintain the license to the IP. Technology transfer offices will often be able to defray the costs for patent filing and other fees so that a newly formed company can have an adequate amount of time to raise funds in order to establish a company. Because of these financial investments, the institution will need to reap benefits of forming the company. Licensing terms vary greatly amongst technology transfer offices; however, entrepreneurs can expect to pay a licensing maintenance fee and negotiate various terms, including milestone payments (clinical studies, fundraising, commercialization expectations, etc.), royalty rates, and the percentage allocated to a university or institution in the event of an acquisition. Some resident entrepreneurs may find themselves in a position where a disclosure is not required; however, this situation is rare, and the author encourages resident entrepreneurs to refer to their experienced team members regarding generation and assignment of the IP.

DE-RISKING YOUR IDEA WITH BUSINESS CANVASING AND DATA GENERATION THROUGH APPROVED STUDIES

With a need defined, a well-vetted solution in place, early prototypes (device or software) at least developed in principle, the idea protected with patent filings, and market opportunity fully explored (target market, serviceable addressable market, and total addressable market), the resident entrepreneur is well served by utilizing tools to organize the business in a blueprint fashion. A useful tool is the "business model canvas" described by Steve Blank and Bob Dork, which can be modified to include the clinical need (Fig. 19.1). The business model canvas should undergo multiple iterations as the team continues to discuss their

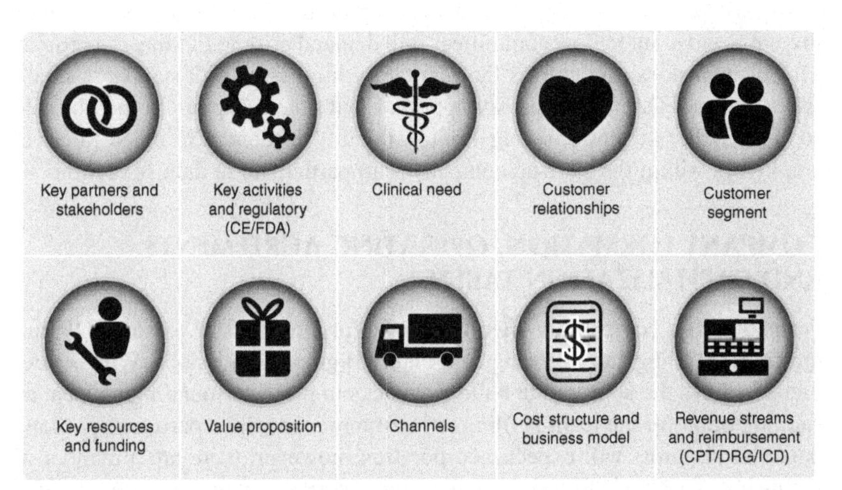

FIG. 19.1 Customized business model canvas with a focus on health care. *(Modified from Blank SG, Dork B. The startup owner's manual: the step-by-step guide for building a great company. K&S Ranch; 2012. p. 36 [Fig. 2.2 Business Model Canvas].)*

solution and market opportunity with various stakeholders at their respective institutions. The business model canvas is a way to provide organization for the group to better understand which aspects of the eventual business plan have not been addressed while allowing the entrepreneurs the ability to seek answers through interviews with potential stakeholders. The author is a proponent of early stage companies performing 50+ stakeholder interviews to best define what solution the team should build while addressing the unmet needs of their customers.

With rapid prototyping of software and medical devices, resident entrepreneurs are in a unique position within their training programs to seek opportunities to collect data with their devices. While small feasibility studies can demonstrate functionality of a device, more robust studies to collect data from their respective prototypes, institutions, and academic centers will require approval from their institutional review board (IRB) and review by the proper committees. Some IRBs may require 510(k) clearance prior to using prototypes of new devices in human studies even for class I devices. For invasive devices, a physician-sponsored investigational device exemption (IDE) may be required by the IRB in order to test the device, necessitating submission to the FDA for review. For instance, software solutions can expect to be evaluated by IT departments prior to deployment, and medical devices can anticipate analysis by hospital biomedical engineering departments to assess safety. Results from an IDE are then utilized by the company to support a 510(k) submission for regulatory clearance. A number of solutions that are more therapeutic in nature will require multiple stages, often beginning with animal studies. This process can be long and expensive but necessary to establish the safety and efficacy prior to moving into human studies. Companies reaching these hurdles will have to show early proof of concept while raising further funds to generate the data most investors and grant review committees will demand prior to funding the project. This chapter does not address the regulatory hurdles or ISO standards established by the FDA for regulatory approval, but other chapters in the book do so. The author suggests having limited liability insurance in place to cover employees within the start-up company who participate in data collection.

COMPANY FORMATION, OPERATING AGREEMENTS, AND CAPITALIZATION TABLES

Formation of a company is often necessary for a number or reasons: (1) fundraising, (2) filing and licensing of IP, and (3) grant submissions. When entrepreneurs have the appropriate building blocks in place surrounding an idea, an exciting point for the team is filing to incorporate or form a partnership. Many potential investors will expect incorporation; however, there are a number of arguments to be made for filing for formation of the company in their respective state or in Delaware, the type of partnership (i.e., LLC, S-Corp, C-Corp, etc.), and the impact on tax filings. Operating agreements are necessary among teams, because team members may leave, and ownership of the company then changes. Vesting schedules and capitalization schemes should be generated with operating agreements so that team members can assign ownership and equity percentages given the participation of each member of the team. Most vesting schedules reward the founding members of the team for the amount of time put in to a company. Industry standards for medical devices involve generally a 4-year vesting schedule with new grant assignments each year of participation for employees within the organization along with 1-year cliffs if a teammate were to leave early. Assignment and ownership of the equity can be a point of contention especially with resident entrepreneurs often limited in their participation due to competing demands on time. Despite this, the author is an advocate for resident entrepreneurs to address equity assignments early with all cofounders. Further, university law clinics and online software tools and templates can be extremely beneficial for resident entrepreneurs to lay the foundation of their company with executed operating agreements amongst cofounders. Failure to establish an operating agreement and capitalization table early in the formation of the company can lead to disputes over ownership when teammates move on to other opportunities. These disputes can lead to failure and dismantling of the company. When done in the right way, an operating agreement can establish a healthy culture and foundation for a company as they move toward commercialization of their innovation or device.

FUNDING AND FURTHERING INNOVATIVE IDEAS IN HEALTH CARE

Perhaps the most challenging aspect of commercializing an idea is garnering resources to advance the company. Medical device companies and health care IT start-ups face a challenging environment for fundraising given the added gauntlet of obtaining regulatory approval (i.e., 510(k) clearance or CE mark) and assignment of a CPT code for reimbursement through third-party payers. Health care IT is often seen as more easily scalable by investors, with fewer hurdles from a reimbursement aspect due to the lack of direct clinical contact with patients. Early on in formation of the company, a number of avenues exist for resident entrepreneurs to fund their ideas. Family, friends, and angel investors remain a good source of capital so that team members can pursue full-time positions within the company; however, these dilutional investments require term sheets which require legal review. Investments in this preseed or seed stage of a company can be equity investments or debt vehicles, including convertible notes, loans, and Simple Agreement for Future Equity (SAFE) notes; SAFE notes do have some downsides so the entrepreneur will really need to do the math of the capitalization table [2].

A number of health care accelerator programs have emerged over the past several years which are also a potential avenue for helping companies commercialize their ideas. A number of these accelerators provide space, cash, and resources in exchange for equity (e.g., $20,000–$100,000 of operating capital in exchange for a 5%–8% equity stake in the company). Examples of health care accelerators include Cedars Sinai Techstars (Los Angeles, CA), Healthtech Wildcatters (Dallas, TX), TMCX (Houston, TX), J-Labs Johnson and Johnson (Houston, TX), MATTER (Chicago, IL), Boomtown (Boulder, CO), Innovation Depot (Birmingham, AL), and DreamIt Ventures (multiple locations). There are also examples of nondilutional capital opportunities for companies, including competitions on health care-focused business plans and government grant opportunities; the National Science Foundation (NSF) Small Business Innovation Research (SBIR) grant awards are a perfect opportunity for resident entrepreneurs to further fund their companies, with the goal of commercializing their technology solutions that may be deemed "too risky" by the more typical investors. Nondilutional phase I grant awards of $225,000 and phase II grant awards of $775,000 are available through an extensive review process. Companies that receive phase II grant awards for commercialization of their technology are in a unique position to capitalize on external investments with matching awards granted for private investments (50% match for external investments greater than $100,000). Phase I companies are required to participate in an informational boot camp which helps entrepreneurs to navigate the process of customer discovery. The National Institutes of Health (NIH) also offer SBIR grants for translational research and commercialization of technology. In addition, institutional opportunities may exist for young investigator awards at the residenťs respective program. These awards are often best explored during a residenťs professional development or research months, so that applications are robust and adequate for review. Ample opportunities for funding exist for resourceful resident entrepreneurs seeking to commercialize their ideas on their pathway toward becoming a successful physician entrepreneur/innovator.

IMPORTANT POINTS

- When a resident is exploring new ideas in health care innovation, the resident entrepreneur needs to evaluate three domains: market size, technical feasibility, and value proposition.
- The ideation phase is best addressed with formation of a cross-functional team to aid company formation, protect the IP, develop software iterations and wire framing to generate a realistic business model, and plan for regulatory approvals and commercialization.
- IP cannot be ignored; Google patent search (https://patents.google.com/) is a great tool to identify existing IP and should be queried similar to PubMed regarding themes and keywords.
- Failure in rapid prototyping of an idea is hard for a resident to accept, but failure must be acknowledged as a possibility to learn from and quickly pivot toward more promising solutions.

- Because a resident entrepreneur is supported by resources and funding through their institution, their inventions must be disclosed appropriately to the technology transfer office which can benefit the entrepreneur.
- The "business model canvas" described by Steve Blank and Bob Dork provides organization for the group to understand which aspects of the business plan have not been addressed.
- IRBs may require 510(k) clearance or a physician-sponsored investigational device exemption (IDE) prior to using prototypes in human studies.
- Establish a clear operating agreement and capitalization table with other members and get their agreement and assure their understanding early in the formation of the company.
- Health care accelerator programs have emerged over the past several years as potential avenues for helping companies commercialize their ideas by providing space, cash, and resources in exchange of equity; examples include Cedars Sinai Techstars (Los Angeles, CA), Healthtech Wildcatters (Dallas, TX), TMCX (Houston, TX), J-Labs Johnson and Johnson (Houston, TX), MATTER (Chicago, IL), Boomtown (Boulder, CO), Innovation Depot (Birmingham, AL), and DreamIt Ventures (multiple locations).

REFERENCES

[1] Schwartz J, et al. Needs-based innovation in cardiovascular medicine: the stanford biodesign process. JACC Basic Transl Sci 2016;1:541–7.

[2] Levensohn P, Krowne A. Why SAFE notes are not safe for entrepreneurs. Crunch Network; 2017.

FURTHER READING

[3] Blank SG, Dorf B. The startup owner's manual: the step-by-step guide for building a great company. K&S Ranch; 2012. p. 36 [Fig 2.2].

Chapter 20

Fostering and Expanding Diversity in the Workforce in Innovation

Priya Kumthekar and Mamta Swaroop
Northwestern University Feinberg School of Medicine, Chicago, IL, United States

THE CURRENT ENTREPRENEURIAL LANDSCAPE

Technology and innovations are permeating into our society, mitigating obstacles in life. The realm of health care is no different with many surges in biotechnology and medical start-ups over the past decade. Substantial changes in health care reform, legislation, and disruptive transformations have led to advances in health care delivery. Within the past decade, the number of venture funding deals in health care is up over 200% [1]. Despite this modernization, there continues to be a blatant lack of female entrepreneurs, specifically in the medical and biotechnology space. The Kaufmann Foundation, a nonprofit organization focused on education and entrepreneurship, showed only 15% of businesses in the biotechnology and high-technology sectors had primary female owners compared to 30% in other sectors. They also found that women founded only 3% of technology firms and 1% of high-tech firms between 2004 and 2007 [2]. These findings suggest a highly male-dominated venture capital industry, which carries with it and insinuates exclusivity preventing both females and possibly other less represented individuals of diversity from readily entering this space. The idea of gender bias in venture capitalist decision-making is not unique to the Kauffman Foundation report. Women entrepreneurs are overall less likely than men to acquire venture capital; a study out of high-tech entrepreneurs in 2001 revealed that only 5% of invested venture capital went to women-owned high-tech or biotechnology firms [3]. Despite our alleged modern society, *masculine* characteristics still seem to yield more *successful* entrepreneurs [4]. With these biases interwoven into our culture, one can only imagine the impact of gender bias on entrepreneurial conquest and investment as well as leadership in the biotech arena.

Medical Innovation. https://doi.org/10.1016/B978-0-12-814926-3.00020-6

THE ROLE OF FEMALE PHYSICIANS IN THE CURRENT LANDSCAPE (OR LACK THEREOF)

The paucity of women in innovation and entrepreneurship is evident; however, an increase in the number of women-led angel funds and incubators focused on women speaks to at least the start of change. How then do we explain the near absence of female physicians in this realm where male physicians have made quite a mark? Female physicians are amazing in and of themselves. Many of these women with families tend to not only run a busy practice but also often bear and nurse their children and tend to be the one in charge of household responsibilities. The authors may sound against their male counterparts, however, on the contrary, even if the male partner has equal responsibility, due to gender schemas, our children, husbands, nannies, and housekeepers usually come to the females when there is crisis. The role of female leadership at home may be controversial, however, many still perceive the female as the "house administrator" through which most decisions regarding children, the home, and the functionality of the individual family members is carried through. This responsibility has sustained over generations despite the evolving role of women in the workplace over the same time span. This overwhelming environment and balancing act combined with certain qualities pervasive amongst females and then topped off with a glass ceiling in the world of innovation and entrepreneurship may explain the dearth of female physicians as innovators and entrepreneurs.

A recent paper out of Harvard Business School showed that between 1990 and 2016, women represented only 8.6% of the entrepreneurial and venture capital labor pool. Of this figure, women in the entrepreneurial ventures of health care made up 10.5%, and women in information technology (IT) made up even less (5.5%) of their respective entrepreneurial pool [5]. These figures are despite a larger representation in the labor pool and a larger representation in educational fields that lead to these careers. In this same time window, women made up roughly 45% of the labor force, while women enrolled in professional fields, such as medicine and law has increased considerably.

EXPLANATION FOR LACK OF FEMALE ENTREPRENEURS

To reveal the underlying causes as to why there is a shortage of female entrepreneurs in medicine could also uncover the persistent mystery of gender gaps in the workplace, both in the United States and around the world. The answer is multifactorial and multicultural, and is seen in almost every country and every industry in varying degrees. Understanding the gender gap in entrepreneurship is multifaceted and complex and extends beyond the world of medicine. The classic arguments for the paucity of female entrepreneurs can be broken down into two main categories: the first is the *supply* theory arguing that there is shortage of women with high-level education, experience, and strategy, and the second theory is that there are ongoing gendered practices that affect the landscape and structural organization of these institutions.

Debunking the first argument that there is a lack of females with high position-appropriate education and experience can be easily disproven. Among first-time applicants to medical school, the number of women increased by 5.3% over 2015, reaching 19,682. In 2016, enrollment in medical school was divided evenly between women (49.8%) and men (50.2%) [6]. Despite the growing number of women in medicine, the number of women leading academic departments is a reflection of the gender gap in medicine. In 2015, of the 2768 department chairpersons, only 437 were female (approximately 16%) [7]. Another recent study looking at data in 2016 showed that only 15% (22/149) of deans and interim deans at medical schools in the United States are women, and the prevalence of women in decanal positions (positions that are dean level) decreases with ascending professional rank ($R^2=0.93$; $P<.05$) [8]. The lack of females in these important leadership positions is representative of the difficulty women have to break the barriers imposed by gender stereotypes and bias. Women in medicine have very different leadership positions than men after surpassing identical academic hurdles [7].

The authors maintain that the social and professional capital of female entrepreneurs is indeed being influenced by gender. These influences can be both internal/self-imposed by the female professional as well as external as a product of societal environment. Often, we tend to focus on societally imposed gender equality; however, an important step in bridging this gap is to understand that many of these challenges to equality are self-imposed. Women, more than men, suffer from the so-called *imposter syndrome*. Imposter syndrome is defined as a concept where one cannot internalize his or her accomplishments and is in constant fear of being exposed as a *fraud*. This syndrome appears to be rampant in females, particularly amongst female professionals. One study looking at the imposter syndrome among medical students showed that approximately 50% of female medical students suffered from some element of this imposter syndrome, which was over double that of the male medical students with the same mindset [9]. Not only does this show the gender difference, but also that this sense of being *fraudulently* successful starts at an early age, making it a mindset that is that much harder to undo. Because confidence and assertiveness is a nearly nonnegotiable trait for an entrepreneur, the risk of this syndrome deterring an entrepreneur-in-the-making is real.

Although the days of men hunting and gathering versus women devoted to childrearing is long past, there remains an ongoing struggle to define traditional family roles within the nuclear unit due to cultural bias but more important, are various internal pressures women place on themselves. Stemming from the gender schema or mostly unconscious hypotheses we have about the different roles of males and females in society, we discount abilities in women and applaud the same in men. We see females as nurturing, communal, and as doing things out of concern for other people. And, we see males as capable of independent action, doing things for a reason, and getting down to the business at hand.

In addition to a paradigm shift culturally, there is also an argument that our education system could do more to inspire female entrepreneurs. While educating women scientists and health care professionals to be outstanding problem solvers and analytical thinkers, there is a lack of exposure to the skills that enable women not only to be field experts but also to be leaders in new business enterprises. Much of this education comes from mentorship as well as recognizing and relating to other female trailblazing entrepreneurs. Historically, there have been fewer female entrepreneurs within reach to serve as role models and mentors, but as we evolve out of our traditional biases and cultural norms, we hope for a snowball effect with more female entrepreneurs in the health care and biotechnology sector available to serve as influencers/mentors/leaders for the next generation of innovators and entrepreneurs.

MODELS OF SUCCESS

Elise Singer—Doximity

Dr. Elise Singer, a family physician practicing geriatric medicine at University of California San Francisco Medical Center, is a delegate to the California Academy of Family Physicians. In the course of her career, she has cofounded several start-ups in health care information technology, including Doximity, the largest professional network for US health care professionals, having more than 70% of physicians as members; moreover, she was Chief Medical Officer of CalHIPSO, the California-wide Regional Extension Center. Clinicians can use Doximity on their iPhone, iPad, Apple Watch, Android device, or computer to connect and collaborate securely with other health care professionals about patient treatment, identify appropriate experts for patient referrals, and manage their careers [10].

Sophia Yen—Pandia Health

Dr. Sophia Yen is a pediatrician out of the Stanford Medical System who works mainly with adolescent medicine. She describes herself as doing "outpatient gynecology" for young adults and adolescents. It was through this work that she identified a need for providing reproductive care to this population. In 2016, she thus founded Pandia Health whose mission is to make women's lives easier by providing a one-stop shop for recurring medications. The doctors who work with Pandia specialize in women's health and have a passion for providing affordable and convenient services for birth control evaluation to our patients. Since its launch in August 2016, Pandia Health has received awards from Girls in Tech and the Palo Alto Social Impact Center. Dr. Yen was named one of the top women entrepreneurs to watch in 2017 by Bustle [11]. She also graduated from both Springboard Enterprises Tech Innovation Hub and Women's Startup Lab's Accelerators. Dr. Yen breaks the mold of the male innovator by relating to women consumers and by relying on her experiences as a mother and a seasoned professional. She takes some of the insecurities of females and turns them into assets in her article in inc.com to empower other women founders.

Julie Silver—Oncology Rehab Partners

Dr. Julie Silver, a rehabilitation physician at the Brigham and Women's Hospital and Massachusetts General Hospital in the Harvard Medical School system, is the founder of the Oncology Rehab Partners. Her idea began in 2003 after being diagnosed and treated for breast cancer. It was then that Dr. Silver realized there was a large gap in helping survivors to heal properly from toxic cancer therapies. In 2009, she created Oncology Rehab Partners and the STAR Program Certification with the mission to improve on survivorship care and to develop a standardized system to deliver optimal survivorship care. Oncology Rehab Partners is an innovative health care company breaking new ground in the field of cancer care. They were the industry's first company to provide hospitals and cancer centers with the tools needed to develop and deliver quality, reimbursable services in oncology rehabilitation to the millions of cancer survivors who suffer from debilitating side effects caused by treatments. In 2012, Oncology Rehab Partners was recognized by Bloomberg/Businessweek as one of the 12 most promising social entrepreneurial companies in the United States. That same year, Oncology Rehab Partners received two Stevie Awards in the business services category—Most Innovative Company and Company of the Year [12]. Dr. Silver's pursuit debunks the unfounded theories that females do not carry the same leadership abilities and entrepreneurial skill. In fact, Dr. Silver was named a *Top Innovator in Medicine* 2012 by The Globe 100. She cofounded the company with Diane Stokes, MBA and as CEO, Dr. Silver has grown it to a company producing about 10 M in revenue a year (owler.com). In 2014, it was partially acquired by McKesson Specialty health, but Dr. Silver remains the CEO and major decision maker for the company.

FUTURE DIRECTIONS

Like any problem, the first step toward a solution is recognizing the challenges and understanding the specific barriers. We have identified multiple potential explanations for the lack of females in the health care and biotechnology start-up sector. These include issues such as overcoming the imposter syndrome, dissolving specific cultural perceptions of female ability and role in business as well as difficulties in obtaining venture capital amongst other barriers.

Looking forward, as a health care and biotech community, we should all aim to improve on multiple environments to encourage and promote female entrepreneurs in this rapidly evolving sector.

The following are action items to encourage this evolution:

- Work on career transitions for females in the health care and biotechnology sectors
- Create growth opportunities within academia and industry for female professionals
- Identify knowledge gaps in gender bias and take action to eliminate knowledge gaps
- Promote the discussion of gender bias among both men and women, specifically in education and the workplace where innovation occurs

The authors believe these steps could improve gender gaps in any industry and in any country around the world. The key to the shortage of females in any field comes down to one simple point: equality in the treatment of men and women. As we rise above this challenging situation, identification and action from businesses and individuals alike are the first few steps toward equalizing the milieu in which we encourage and bring forth the next generations of entrepreneurs and innovators not only by bringing in more women but also by aiming for further diversity as a whole. From a Darwinian perspective, with increased diversity comes more productivity and a greater chance to forward the biotech and medical ecosystems from how they exist today.

IMPORTANT POINTS

- The social and professional capital of female entrepreneurs is influenced by issues in gender which can be both internal/self-imposed by the female professional as well as external as a product of our current societal environment.
- Women more than men suffer from the so-called imposter syndrome, defined as when one cannot internalize his or her accomplishments and is in fear of being exposed as a fraud. Because confidence and assertiveness is a nearly nonnegotiable trait for entrepreneurs, the risk of this syndrome deterring a female entrepreneur-in-the-making is very real.
- There is a lack of exposure of women (and men) to the skills that enable women to be experts in certain fields in medicine and leaders in new business enterprises.
- The first steps toward a solution involve overcoming the imposter syndrome, dissolving specific cultural perceptions of female abilities and roles in business, and addressing difficulties in obtaining venture capital amongst other barriers.

REFERENCES

[1] http://fortune.com/2015/02/19/healthcare-startups-succeed/.
[2] Robb A, Coleman S. Characteristics of new firms: a comparison by gender. The Kauffman Foundation; 2009. Retrieved from: http://www.kauffman.org/uploaded/Files/kfsgender020209.pdf.
[3] Brush C, Carter N, Gatewood E, Greene P, Hart M. Gatekeepers of venture growth: the role and participation of women in the venture capital industry. The Kauffman Foundation; 2006.
[4] Gupta VK, Turbban DB, Bhawe NM. The effect of genderstereotype activation on entrepreneurial intentions. J Appl Psychol 2008;93:1053–61.
[5] Gompers PA, Wang SQ. Diversity in innovation; 2017. Harvard Business School Working Paper, No. 17–067.
[6] http://news.aamc/medical-education/article/women-enrolling-medical-school-10-year-high/.
[7] https://www.aamc.org/download/481206/data/2015table11.pdf.
[8] Schor N. The decanal divide: women in decanal roles at US medical schools. Acad Med 2017; https://doi.org/10.1097/ACM.0000000000001863.
[9] Villwock JA, Sobin LB, Koester LA, Harris TM. Imposter syndrome and burnout amongst American medical students: a pilot study. Int J Med Educ 2016;7:364–9.
[10] https://www.linkedin.com/in/elisesinger.
[11] https://www.inc.com/springboard/four-undervalued-characteristics-that-lead-to-better-startups.html.
[12] http://oncologyrehabpartners.com.

Chapter 21

Preparing America's Entrepreneurial Workforce: Reinventing the Medical Curriculum

Mark S. Cohen and Seth Klapman
University of Michigan, Ann Arbor, MI, United States

INTRODUCTION

Medical school and residency educational curricula continue to be in a state of flux, as trainees need to process more information than ever in a limited timeframe and accrediting bodies are increasingly rigid about training requirements. Many schools and programs have adapted to this information overload by condensing some topics and making information more high yield to learners. In many cases, the traditional classroom has been transferred or inverted to store more content in digital media where learning how to access the right information has become a key tool to aid physicians with diagnosis and treatment for patients. Likewise, experiential learning is increasingly quantitative and required by accrediting bodies and, therefore, time for exploratory learning is limited. As such, implementing new curricular opportunities can be a challenge given the limited bandwidth of current medical learners.

In today's medical environment, physicians must learn to utilize and evaluate new technologies for their practices, placing them on the frontline of the workforce interfacing with this vast influx of new medical technology. While it is one thing to be able to evaluate the utility and benefit of a new technology, it is an altogether different skillset to troubleshoot problems in medical practice and create innovative solutions to improve patient care. Unfortunately, the vast majority of residents and medical students have no formal educational training on the innovation process and, therefore, in taking good ideas successfully into clinical and commercial implementation. To address this challenge, some institutions have created educational resources and curricular opportunities in innovation and entrepreneurship to enhance opportunities for trainees to engage

Medical Innovation. https://doi.org/10.1016/B978-0-12-814926-3.00021-8

in this effort and enhance their training. One particularly robust effort in this regard is the creation of a cocurricular pathway of excellence (PoE) in Innovation and Entrepreneurship (I&E) for medical students at the University of Michigan.

THE MEDICAL SCHOOL PATHWAY OF EXCELLENCE IN INNOVATION AND ENTREPRENEURSHIP

Recognizing the need for medical students to become acquainted with creating new medical technologies and to better interface with the plethora of medical innovations emerging into the market and premarket studies each year, it becomes paramount for academic medical centers (AMCs) and medical schools to offer more formalized training in this area. Only through commitment of resources, development of novel curricula, and better engagement between academia and industry can a real culture of innovation/entrepreneurship develop. Such development would provide real impact, returns on investment, and improve the quality of technologies being offered to patients. In order to meet the needs of learners working within the limited constraints of today's medical education, resources for innovation and educational opportunities need to be adaptable for trainees to have meaningful participation and engagement. Traditional lectures often fall short in engaging millennial learners who have very different learning styles compared to other generations. As such, experiential learning, digital media platforms, and reversed, or flipped, classroom opportunities should be considered to optimize engagement and interest from such a diverse group. One solution that was adopted at the University of Michigan and has been highly successful in this teaching environment is the creation of a team and network of dedicated individuals who can work with all facets of academic learners, from students to senior faculty and chairs, to provide a variety of educational programs, including online modules, formal courses, workshops, train-the-trainer models, and real-world coaching and mentoring expertise from successful entrepreneurs, senior industry leaders, and venture capital fund managers.

In addition to the foundational education in navigating the business development canvas, schools must also commit financial resources toward trainee and project development, de-risking, and acceleration. These resources are critical to establishing and sustaining a culture of innovation as well as providing an output mechanism to move technologies out of the university to where they can have meaningful clinical impact. By themselves, small seed grants to fund innovation projects without institutional resources to move projects forward, identify killer (GO/NO-GO) experiments and meet milestones, often are ineffective. It is equally important to have a commitment from offices of technology transfer to provide key diligence on patentability, regulatory issues, and the competitive landscape to help identify barriers, pivot earlier, or fail before new technologies are infused with capital that will not be able to advance the technology. The last resource that is often the hardest to allocate at major AMCs is time for faculty, trainees, or students to engage fully in efforts in innovation, discovery, and development. While several academic centers are finally acknowledging faculty involvement in innovation as part of the promotion and tenure process for faculty and an equally important mission of academic medicine in parallel with research, teaching, and clinical care, most have not adopted this as a means to incentivize, recognize, or promote faculty or students. This is equally true with medical students, graduate students, and resident trainees, where time is limited for these innovation activities which have been marginalized as "extracurricular" at best at most centers. The importance of innovation to medicine's future, and prerequisite education, cultivation and deployment of resources to support medical students, residents and faculty in this I&E initiative as a critical but underserved mission was recognized by myself and several other champions at our medical center. We embarked on the programmatic development of an I&E curriculum for medical learners that recognizes unique challenges for medical technologies and industries compared to other business sectors.

In the spring of 2015, a survey was conducted to the current first-year medical school class of 170 students at the University of Michigan asking the question of whether students, if offered the chance, would participate in a 4-year cocurricular pathway in I&E to develop their ideas and learn more about this area for their future careers. Thirty-five percent responded they would participate in such a path, and this lead to the development of the first "PoE in I&E" being created and approved by the medical school curriculum committee for implementation in the fall of 2015.

As with many large AMCs, champions or a center of I&E were already in existence on campus, and each school had created its own academic approach to I&E to facilitate engagement with students and faculty. As such, one main goal of the PoE in I&E was not to replicate or duplicate work or programs already being implemented on campus, but rather to incorporate relevant and high-yield content, infrastructure, and opportunities from these other programs. In addition, the goal was also to add to the unique aspects of these existing programs and help allay some of the challenges that medical technologies and industries navigate in order to get to patients and the market to create meaningful impact.

Bringing together faculty champions in innovation from around campus, including the business school, engineering school, the medical school (especially the Fast Forward Medical Innovation program: a medical school-funded initiative to engage and develop innovation among faculty at the medical center), as well as entrepreneurs and mentors in residence, the PoE in I&E was approved in the fall of 2015 and in its inaugural year, admitted 31 first-year medical students and 9 students from the second- or third-year medical school classes.

The mission of the PoE in I&E is to provide our physicians in training the resources, perspective, and exposure they need to incorporate innovative strategies and tools that can improve the accessibility, quality, and equity of medical care. In doing so, the PoE strives to develop medical students who can discover, examine, and address patient needs creatively through medical innovation and entrepreneurial opportunities and to provide them the tools to explore the transformational role physician-innovators can have on the future of health care [1].

In constructing the educational/curricular portion of the PoE, many moving parts of the complex medical school curriculum had to be traversed. The I&E curriculum needed to have flexibility to adapt to an ever-changing medical school educational environment; we recognized this adaptability was highly important to build into the infrastructure so that it could be utilized effectively by learners whose time was limited and whose priority was their medical school courses (and in today's world, their grades). As such, we learned that a standard, semester/year-long, didactic course, while traditional for many business schools, was not effective or well received by students or faculty/administration in the current learning environment of medical school. This observation had become evident from the results and enthusiasm of a year-long course that was developed for residents and fellows through the Fast Forward Medical Innovation (FFMI) program called the PACE (Program to Accelerate Commercialization Education) course. This 34-week evening course offered over two semesters was constructed to address the needs of medical innovators and provide them with high-yield, interactive, case-based topics traversing the canvas of business development built on the principles of I-Corps training provided to small businesses from the National Institute of Health and National Science Foundation. We enhanced this framework with content specific and unique for medical technologies, such as hospital economics and purchasing, CMS approval, DRG and CPT coding, and reimbursement strategies (Table 21.1).

The course was attended initially by the PoE medical students, but it became a real challenge to attend regularly given the restraints on their diverse class and time commitments. As such, several adaptations were created over the following year to enhance educational content and offerings. These included shortened, high-yield versions of the PACE course focused on customer discovery, development of value proposition, and understanding the key components and importance of a compelling pitch (Fast-PACE and First-PACE) courses: (https://innovation.medicine.umich.edu/education). Additionally, lectures were converted to online videos enhanced with self-assessment modules for students to take advantage of self-directed learning. In addition to the PACE course(s) and online modules, the medical school committed several 1–2h blocks of dedicated pathway time built into the curriculum throughout the academic year to allow the interface of the content of the pathway with students. During this time, a more engaging, reversed classroom was created to showcase real business case scenarios with entrepreneurs talking about their challenges and how they navigated their environment to become successful or discussing their failures and why they failed. As students complete their educational hours

TABLE 21.1 Weekly Innovation Course Lectures Covered in PACE

- Intro to Innovation
- Where Do Good Ideas Come From?
- Clinical Decision Tree and Ecosystem Mapping
- Concept Brainstorming
- Review: Project Incubation
- Project Formation
- Testing Problems Versus Solutions
- Customer/Value Proposition
- Workshop Intro to Reimbursement
- Reimbursement Workshop
- Market Sizing
- Market Sizing Workshop
- Rule of Thumb for a Medical Business
- Intro to Hospital Economics
- Hospital Economics/Purchasing Decisions Workshop
- Intro to Regulatory Pathways
- Intro to Intellectual Property
- Prior Art Searches
- Alternative Protection
- Freedom to Operate
- Developing an IP Strategy
- Developing a Regulatory Strategy
- Clinical Trial Design
- Clinical Trail Workshop
- Medical Device/IT Development
- Solution Brainstorming & Early Prototyping
- Preapproval Regulatory Requirements
- Running a Company/Practice: Lawyers
- Running a Company/Practice: HR
- Funding a Start-up
- Start-ups—Equity & Stock Options 101
- Concept Pitch
- Concept Pitch Practice
- Pitch Presentations

or modules, they are each paired with faculty mentors and innovation advisors to help them cultivate their own ideas in innovation as well as work together on teams with other medical students or graduate students in engineering, business, and law. Beyond the pathway offerings, the medical school also has a number of student-lead extracurricular groups that offer greater opportunity to explore. The Michigan Innovation Group (MIG) invites innovative leaders from across campus and private industry to engage students in lunch talks and roundtable discussions. In a partnership between the Michigan Surgery Interest Group (MIG) and other innovation organizations across the UM campus, the PoE students oversaw the creation of a medical student competition program fashioned after the television show Shark Tank. This important initiative rapidly became a successful, 8-month incubator program for teams of medical students to create innovative ideas in health care around surgical problems. This "Shark Tank" culminated in a finale in which teams competed in front of seasoned venture capitalists and entrepreneurs for Medical School-funded seed grants. Simultaneously, the medical school fully supported the founding of a Michigan chapter of SLING Health (formerly Idea Labs), a national student-led organization developed originally at Washington University designed to focus on combining medical, engineering, business, and law students into multidisciplinary teams to address health care solutions through biodesign. By providing resources, funding, and mentorship for this Michigan chapter , the medical school was integral to the founding year of the chapter in 2016 with six successful teams completing the 9-month program (Fall 2016—Spring 2017). Participation in the chapter includes expert coaching and mentoring, engineering design review evaluations, funding for prototype development, and a DEMO-Day competition in a format styled after Shark Tank, where the winning teams are sponsored to attend the national SLING Health competition.

Once learners in the PoE have mastered a fundamental core curriculum to bring them up to speed regarding how to generate impactful medical innovations and de-risk these ideas in an iterative fashion toward commercialization and clinical impact, each student then works on developing a mentored capstone project through a deeper-dive engagement. Each capstone is unique and designed by the student to explore more fully an innovation of their interest related to its commercialization and its pathway of business development.

In order to get ideas for their capstone if they do not have one or two candidate ideas already initiated, students are encouraged to engage with faculty mentors on innovation projects in their field of interest or participate in internships and externships in innovation that are geared to provide mentorship and real-world innovation/entrepreneurship experience. Activities available to the first cohort of PoE students include 8–12 week internships at a venture capital firm in the area working on company evaluation for investing as well as following due-diligence on invested companies to ensure they are meeting their milestones and growth targets. Other opportunities have included interning at a local start-up, working with engineering students on a biodesign course project, and internships in industry with partners established through the university. Together, these experiences have been invaluable for entrepreneurial development and have helped students with innovative ideas spin out start-up companies and compete successfully for follow-on funding. This ecosystem of innovation involving mentor-mentee relationships within the university has been a critical component to developing a culture of innovation among medical students, residents, and new junior faculty. In order for these younger innovators to be successful and navigate their business development ideas and technologies toward the clinic and patient impact, they have to have great mentorship along the way.

To create a diverse group of faculty innovators to serve in this capacity has been another critical mission of the Medical School, and the FFMI program has taken a lead role in this process, creating numerous courses and programs of idea development for faculty to help them move technologies forward, de-risk them, and provide a strong value proposition for those ideas to move into licensing deals or start-up companies. With every department of the medical school participating in these programs, students have the great benefit of working with a diverse group of faculty innovators as well as a large number of medical innovation projects. This approach has helped to create an ideal culture and environment to train the next generation of medical innovators for success.

THE SURGICAL INNOVATION AND ENTREPRENEURIAL DEVELOPMENT PROGRAM

An example of this type of unique training for faculty occurred in the Department of Surgery. The Department of Surgery at the University of Michigan consists of seven sections with 140 faculty and performs more than 30,000 operations annually; in addition, our department has been ranked in the top five among surgery departments in NIH funding each year for the past decade. To address the challenges of training busy surgery faculty in I&E, the FFMI team developed a tailored, 9-month program for faculty called the Surgery Innovation and Entrepreneurship Development Program (SIEDP) [2]. By providing an educational and mentoring program for faculty in I&E, the SIEDP helped to generate innovative approaches to improve the department in delivery of clinical care, education, research, and outreach. In the first session, participants brainstormed ideas and then prioritized these ideas from 60 down to 10 top innovations. These 10 were then voted on at a department-wide faculty retreat, resulting in the top four innovations advancing through the course in teams of three to four faculty each. Each program meeting (see Agenda in Table 21.2) was used as a working session for teams to advance their projects, resulting in minimal required work and schedule disruption outside of the program. Under the guidance of a dedicated Lead Instructor with deep expertise in biomedical technology development and entrepreneurship, augmented by guest lectures from the Medical School, College of Engineering, Law School, School

TABLE 21.2 Surgery Innovation and Entrepreneurship Development Program (SIEDP) Program Agenda [2]

Session 1
• Delivering a Business Pitch • Trends in Medical Innovation • Ideation Session
Session 2
• Team Formation • Value Proposition Analysis • Stakeholder Maps
Session 3
• Commercialization Case Study • Stakeholder Map Presentations • What Is Customer Discovery?
Session 4
• Risk Analysis • The Executive Summary • Intellectual Property and Start-ups
Session 5
• Med/Engineering Collaborations

of Information, Tech Transfer, and interactions with FDA regulatory leadership, private industry participants, and venture capital managing partners, the individuals, and project teams worked to develop a comprehensive business case and a viable commercial pathway for their innovations. Teams were coached on business development through hypothesis-testing and validation through a process called "customer discovery," an interviewing technique developed by Steve Blank for his Lean Launchpad course at Stanford University [3,4]. This experience provided all learners with exposure to strategies for commercialization, innovation, and entrepreneurship that can be used throughout their careers to creatively address patient needs creatively via a rigorous and methodic process of medical innovation and commercialization [2].

Building on successful faculty programs to develop lead faculty champions in I&E across every department in the medical school has allowed the next generation of medical innovators, including medical and graduate students, residents, fellows, and new faculty, to develop a robust group of field-specific mentors to coach them and help their projects move forward in the local ecosystem.

EVALUATING RETURN ON INVESTMENT

By having outstanding resources including great faculty mentors, high-value educational programs, funding for seed projects, time to participate in internships/externships, and commitment from students to participate in the PoE, a sound foundation for success has been created. But how can success be measured for a program and culture change? This is equally important for institutions to delve into and understand what their goals of success should be and what metrics will they use to determine if that success has been achieved. For our PoE program, we knew that success could not be measured in just financial revenues, because this type of program is much more than a line-item on a budget. Instead we looked at three different parameters to grade productivity from our program. The first was engagement. In year one, 40 students joined the PoE, and by year two, this number increased to 60 students. Additionally, our SLING Health biodesign teams and the MIG Shark Tank groups engaged another 80–100 students in innovation projects spanning four schools on campus. Each of those student projects had faculty mentors and coaches engaging over 100 faculty across every department in the medical center in innovation efforts. These new programs were recognized for two national innovation finalist awards.

The second level of return on investment (ROI) was in programs. Through the PoE and FFMI, four new educational courses were developed (Full PACE, Fast PACE, and First PACE along with online I&E modules). Additionally, we have held five Shark Tank competitions and have created an annual, statewide competition for medical and graduate students sponsored, in part, by the Michigan Economic Development Corporation (MEDC). Our students have participated in I-Corps training, as well as competed successfully to participate in several incubator and accelerator programs both regionally and nationally, including VentureWell, Ann Arbor SPARK, TechTown Detroit, and others. We have also engaged donors, alumni, and several departments in the medical school and engineering school to sponsor funding for student innovation teams to seed projects or help create key prototypes needed for additional investor funding. One key but unexpected finding as a result of this educational exposure was a substantial increase in the number of dual-degree MD/MBA medical students who are now interested in innovation as part of their future medical careers. For the past decade, the number of MD/MBA students in each class varied from three to six, but this year, 15 students applied, demonstrating the important impact of training medical students in I&E on the future medical workforce.

Finally, the last metric we evaluated was output. In just the first year of the program, six new medical student start-ups were created from student innovations, with three having fully functional products that already have been implemented clinically in several developing countries, thereby highlighting the importance of social entrepreneurism within medical innovation. Additionally, all of the medical student start-ups have now successfully competed for and attained follow-on funding to move their companies forward, and a few of the founders have taken an additional year, apart from the formal medical school classes, to run their new company and move it forward. The program has also attracted a substantial number of alumni and donors who are being engaged by development to create lasting funding and opportunities for students to utilize and thrive in this unique program and environment of I&E. Together, this will help create the leaders and the best in medical innovation and provide a strong cohort of well-trained physician-innovators who will help lead the ever-changing workforce of medicine and its vastly growing dependence on technology to advance the care of patients.

CONCLUSION

While the past culture of academic medicine has not fostered academic innovators due to restricted resources and arguably a naïve lack of recognition related to academic advancement, the culture is changing, and many AMCs today are investing in innovation. A key component of this change in culture has come from deployment of critical resources to bolster innovation efforts. These resources include mentorship, coaching, educational programs, seed financing, and time and effort allotted to faculty (and now students as well) to explore their innovative ideas and be able to use their innovation efforts toward academic advancement. These efforts by AMCs are developing forward-thinking academic scientists and clinicians who are well versed in the fundamentals and importance of creating a meaningful value proposition, understanding the market and the customer, navigating the regulatory constraints of the current environment, and establishing the financial framework required to move a good medical innovation successfully to impact patients [1].

In order to prepare America's entrepreneurial work force in medical technology, there is a need for creating a culture that not only accepts medical innovators and entrepreneurs as valued members of AMCs, but also encourages them to engage and be successful through an infrastructure rich in the resources of education, mentoring, and funding. Programs like the PoE in I&E at the University of Michigan and other institutions are paving the way for the next generation of medical and surgical innovators who will advance the field exponentially [1].

We have arrived at a historic inflection point where a major determinant of whether a new biomedical technology will move forward into the clinic is based on cost and reimbursement. Navigating this environment will require a thorough understanding of the economics of health care, where university faculty members who specialize in health policy and economics will be an important resource for tomorrow's medical innovators [5]. Selected and interested students completing educational programs in I&E such as those in the PoE need to be trained in the tools to develop and de-risk their ideas for real clinical benefit [1]. By investing in resources and meaningful educational opportunities, AMCs can develop and equip creative physician-innovators, who can have a substantive and enlightened impact on the future of patient care.

IMPORTANT POINTS

- To meet the needs of medical students, graduate students, and physicians in training, resources for education and engagement in innovation and entrepreneurship need to be made available in academic medical centers.
- When first-year medical students at the University of Michigan were asked if they would be interested and participate in a 4-year cocurricular program in Innovation and Excellence, 35% wanted to participate in the Pathway of Excellence in Innovation and Entrepreneurship (PoE in I&E).
- The mission of this PoE in I&E (and similar programs in other universities) is to give physicians in training the education, resources, perspective, and exposure to explore a transformational role for potential physician-innovators.
- The PoE in I&E involves a reversed-classroom approach to showcase real business case sessions with entrepreneurs as well as lectures, some of which were converted to online videos to enhance self-learning, and self-assessment modules to enhance engagement.
- The medical school also committed several 1–2 h blocks of dedicated pathway time built into the curriculum throughout the academic year.
- After completing their educational curriculum, medical students are paired with faculty mentors and innovation advisors to work together on teams with other medical or graduate students in engineering, business, and law
- The Michigan Innovation Group (MIG) supplements this program by inviting innovative leaders from across campus and private industry to engage students in lunch talks and roundtable discussions.
- After mastering the fundamental core curriculum, students then develop a mentored capstone project designed to learn about commercialization and the pathway of business development.
- Other opportunities include interning at a local start-up, working with engineering students on a biodesign course project, and internships in industry with partners established through the university.

ACKNOWLEDGMENTS

For all their efforts, vision, and collaboration in the development of these innovation programs for medical students, residents, and surgery faculty, I would like to acknowledge the University of Michigan Fast Forward Medical Innovation Team especially Kevin Ward, MD, Constance Chang, MBA, Jon Servoss, MEd., and Margarita Hernandez, PhD; the Office of Sponsored Research; David Olson, PhD from the Office of Technology Transfer; Jonathan Fay, PhD, Interim

Executive Director of *Michigan* Engineering's Center for Entrepreneurship; Senior Associate Dean for Medical Education Joseph Kolars, MD, Associate Dean for Medical Education Raj Mangrulkar, MD, Paths of Excellence & Leadership Co-Directors Heather Wagenshutz, MA, MBA and Eric Skye MD, and Michael W. Mulholland, MD, PhD, the Frederick A. Coller Distinguished Professor and Chair of Department of Surgery at the University of Michigan. Funding for this program and manuscript was provided in part by the Department of Surgery and the University of Michigan School of Medicine.

REFERENCES

[1] Cohen MS. Enhancing surgical innovation through a specialized medical school pathway of excellence in innovation and entrepreneurship: lessons learned and opportunities for the future. Surgery 2017;162(5):989–93.

[2] Servoss J, Chang C, Olson D, Ward KR, Mulholland MW, Cohen MS. The Surgery Innovation and Entrepreneurship Development Program (SIEDP): an experiential learning program for surgery faculty to ideate and implement innovations in health care. J Surg Educ 2017. https://doi.org/10.1016/j.jsurg.2017.09.017. [Epub ahead of print].

[3] The National Science Foundation. Innovation corps commemorated [press release]. Retrieved from: http://www.nsf.gov/news/news_summ.jsp?cntn_id=124935; 2012.

[4] Blank SG, Dorf B. The startup owner's manual: the step-by-step guide for building a great company. Pescadero, CA: K&S Ranch, Inc.; 2012.

[5] Yock PG, Brinton TJ, Zenios SA. Teaching biomedical technology innovation as a discipline. Sci Transl Med 2011;3(92):1–6.

Chapter 22

A Dean's Perspective on Entrepreneurship in the University

Kevin E. Behrns and Andy Hayden
Saint Louis University School of Medicine, St. Louis, MO, United States

INTRODUCTION

A university and a school of medicine serve four major functions: (1) education of learners, (2) creation of new knowledge, (3) transfer of new knowledge or innovation into a product or business (entrepreneurship), and (4) service to the community. This chapter will focus on the perspective of the role of a Dean at a medical school at promoting innovation and the subsequent creation and development of a product through established models to create a business or entrepreneurship.

The creation of new knowledge and the transfer of such is the fabric of major research universities. All colleges and universities teach and provide service to the community, but major research universities and in particular schools of medicine distinguish themselves through research, innovation, and more recently via entrepreneurship. For decades, schools of medicine have strived arduously to create new knowledge primarily through basic science departments that function to address fundamental issues in biochemistry, physiology, pharmacology, and the like. These efforts have produced many groundbreaking discoveries that served as the basis for innovation in drug design and discovery, for example. In the past two decades, basic science research has remained strong but, in many institutions, the focus has shifted to clinical research that will directly impact patient care. This transition in the emphasis of research has been encouraged by funding agencies and the need to bring new treatments to patients in a shorter period of time. Faculty members engaged in this type of research soon realized that if they could discover a new treatment, they could test the product in patients and work with experts in business development to build on the concept, commercialize the product, and then take it to market. On occasion, this process of discovery and entrepreneurship resulted in

Medical Innovation. https://doi.org/10.1016/B978-0-12-814926-3.00022-X

lucrative financial rewards for the university and also on occasion for the researcher. This relatively new business model, inclusive of physicians, elevated entrepreneurship in academia.

The aim of this chapter is to familiarize the reader with the current state of entrepreneurship at a major research university, and, in particular, a school of medicine. In addition, we will discuss how this important aspect of entrepreneurship influences the missions of academia in clinical care, research, and education. Finally, we will review how to create an environment of entrepreneurship and how such a program can foster growth in universities' communities.

ENTREPRENEURSHIP AT A UNIVERSITY

As the United States shifted from a manufacturing economy to a knowledge economy, major research universities became an engine for innovation, because new knowledge was created and held by their faculty. As faculty interest in entrepreneurship increased, so did the interest of the students with whom they work. And as students increased their involvement in innovation, their interest in the business aspects related to product development and commercialization were piqued, and they desired further knowledge about the entirety of the entrepreneurship model. Thus, a major in (or formal study of) entrepreneurship for undergraduate students was developed in some universities. Today, these students often live in communities or dormitories that house only students enrolled in this or related majors, and often their classes are held in innovation hubs or "incubators" that may be somewhat secluded and/or remote from the medical school.

An education in entrepreneurship is typically housed in the school of business in the university and includes course work or experiences in resource management (including human resources), business plan development, organizational behavior, finance/capitalization, ethics, negotiation, leadership, organizational failure, and change and transformation, often with a hands-on experiential internship. Undergraduate students at major universities majoring in entrepreneurship learn from a diversity of faculty members across the university in the business school, STEM (science, technology, engineering, and math) disciplines, and ethics, depending on their primary field of interest.

Aside from teaching entrepreneurship, a minority of faculty, often STEM or medical research faculty, at major research universities, especially those affiliated with a medical school, are also intimately involved in entrepreneurship. Generally, these faculty members have discoveries that can be commercialized. Frequently, the discoveries occur in university laboratories or clinics, whereas the commercialization phase is outsourced to organizations that are more adept at product development. For the faculty members, this often creates the dilemma of having a full-time faculty appointment but desiring a position as a chief scientific officer in their new start-up business; this dilemma can result in a conflict with their university appointment (see Chapters 13 and 14).

The financial rewards for successfully developing a product that originated from basic science work at a university may be quite generous. Both the university and the faculty member may experience quite substantial financial gain. Most universities have well-prescribed formulae that distribute rights and/or royalties to the university and the inventor. Some products developed from universities have led to major financial windfalls that positively changed the reputation and even the direction of the university.

Currently, the traditional models of financial sustainability at major universities are under siege, because the competition to attract students is fierce, and price points are under considerable pressure as net tuition decreases markedly. As a result, more universities are focusing on converting knowledge into products, not only to enhance their reputation but also as a revenue stream. In this current age of increasing knowledge, innovation and entrepreneurship have become major components of strategic plans of some universities and medical schools.

ENTREPRENEURSHIP IN A MEDICAL SCHOOL

Entrepreneurship in a school of medicine can positively influence the clinical, research, and education missions. Furthermore, in a well-designed program, entrepreneurship may result in financial rewards that can further enhance research and/or education programs. Increasingly, students and trainees seek entrepreneurial education and view entrepreneurship as a career opportunity during or after medical school or clinical or research training. Below, we focus on the role of entrepreneurship on the missions of a medical school in the fields of clinical work, research, and education.

Entrepreneurship and the Clinical Mission

Entrepreneurship has been aligned tightly with the clinical missions for years. For several decades, a readily visible connection between clinical care and innovation was the production of medical devices, particularly in specialties like cardiovascular surgery. Cardiovascular surgeons worked closely with bioengineers to develop the heart-lung machine, biomechanical heart valves, ventricular assist devices, and numerous other products. These products were developed in response to vexing clinical problems that, when unresolved, often resulted in immediate death or a markedly shortened life span. In more recent years, the pace of product development has continued to be brisk with the emergence of laparoscopic, endovascular, and minimally invasive surgery allowing procedures such as transcatheter aortic valve replacement (TAVR). These evolving fields continue to provide interesting research questions that must be addressed via basic research, clinical trials, and product development.

Thus, a close working relationship between an entrepreneurial clinician and device or start-up companies should and needs to be encouraged. Recently, much interest and debate has arisen from academe and certain interests of the government and the press on potential conflicts of interest that may exist between a clinical entrepreneur and a for-profit device company, but these relationships between academe and industry are essential for continued growth and ingenuity for innovative approaches to unresolved clinical problems or improved, less-risky approaches to diseases or conditions.

Likewise, clinical translational scientists have been engaged in drug development for quite some time. These highly trained and observant scientists have thoughtfully recognized the need for innovative drug development, and they have worked closely with industry to develop new pharmaceutical approaches to cancer, heart disease, and many other acute and chronic diseases. A current example is the burgeoning therapy for blood cancers, CAR-T (chimeric antigen receptor therapy). This immune-modulating approach is rooted in developments from Steven A. Rosenberg's Tumor Immunology Laboratory at the National Institutes of Health Center for Cancer Research. Dr. Rosenberg has spent his illustrious career studying cancer immunology, and he has made numerous seminal contributions to the field. His work paved the way for Carl Jung, a clinical translational scientist at the University of Pennsylvania, to develop the adoptive transfer of one's own T-cells with tumor-specific receptors that can recognize and kill cancer cells without destroying noncancer cells. This work came largely from dedicated academic clinicians who partnered with industry to create a product that could be mass produced to accommodate patient need. This therapy received approval recently from the United States Food and Drug Administration. Novartis, a market leader in CAR-T, and Gilead, who just purchased the start-up company Kite Pharma for $11.9 billion, are leading the industry effort for this product. Of note, Kite Pharma was started by the urologic oncologist, Dr. Arie Belldegrun, head of the Division of Urologic Oncology at UCLA, who did his research training at the NIH. These stories are examples of the keen insights from well-trained clinicians who applied their clinical observations/insight to the entrepreneurial development of a groundbreaking therapy that likely will cure thousands of patients. Clearly, this story is a win for patients, scientists, industry, and society.

Entrepreneurship in the Research Mission

As noted in the introduction, the mission of a university and medical school is to educate students, create new knowledge, and transfer knowledge into direct benefits for the community. The research mission at a university is the embodiment of *all* these tasks. Traditionally, when one considers creation and application of new knowledge from a medical school, one thinks of a basic science laboratory that is discovering a new cellular or genomic-based pathway, a new biomarker in disease, or a new line of effective drugs to treat difficult diseases.

Increasingly, however, discovery at a school of medicine involves disciplines like engineering applications of nanotechnology, biomechanical devices, and software applications used to transform existing businesses or work flows usually in the field of medicine and its related professions. Literally, there are few inherent boundaries to the applications of new discoveries at a medical school. Therefore, the research mission is the core of innovation and entrepreneurship.

Over the last several decades, prominent research universities have realized the differentiating capacity of their research enterprise. What distinguishes the best universities is their ability to attract thoughtful faculty and students who thrive on discovery and innovation. Once engaged in a culture of discovery, these faculty members and students make their unique contributions that create a virtuous cycle of creation of new knowledge and the applications of discovery science. When these discoveries are translated into meaningful consumer products, the university benefits financially. Though approximately 90% of drug discoveries come from research and development programs in pharmaceutical companies, a minority of drugs are developed in academic research laboratories. When a blockbuster drug (>$1 billion in annual revenue) results from this work, the medical school and the scientist may realize transformational financial gains. Take for example Northwestern University (NU), with the serendipitous discovery of Lyrica, a drug used to treat epilepsy and fibromyalgia; NU catapulted to the top of the list of research universities with large licensing incomes in 2014. In addition to dollars from licensing, NU has an agreement with Pfizer for royalties yielding about $1.4 billion, which NU used wisely to create an endowment. This endowment fostered the development of the virtuous cycle of research growth, because administrators at NU have used the funds to enhance facilities, recruit stellar faculty, and invest in additional drug discovery programs that may yield further financial gains. Traditionally, research universities have not often been the originators of drug discovery, but with pharmaceutical companies decreasing their investments in research and development, schools of medicine and universities are increasingly seeking opportunities for innovation and entrepreneurship in the field of drug discovery.

Though this section of the chapter has used innovation and entrepreneurship in drug discovery as an example of how universities are increasingly investing in research programs, the investments are much broader and include opportunities in the fields of engineering, computer sciences, physical sciences, and many others. While universities seek to transfer knowledge through innovation and entrepreneurship and are looking increasingly for revenue enhancement through this mechanism, the inherent risks of such a program need to be understood and should be articulated clearly. First, innovation and entrepreneurship require more than just an innovator. Yes, a university and medical school needs a skilled team of administrators who assist the investigator and protect the university and medical school with the many nonscientific components of developing a product (see Chapter 6). Second, the university will often need partnerships with industry and the business world for product development, commercialization, and investment. The opportunity to locate and engage such partners occurs most readily when the university is colocated in a community with a physical innovation hub and/or incubator (see below). Third, the likelihood of hitting a financial windfall from a program in innovation and entrepreneurship is small, and, thus, investments in research programs must be judicious, and administrators should understand that some will fail. These investments should be guided by a multidisciplinary committee with specific expertise in entrepreneurship.

The research mission at a university or medical school is suited uniquely to develop a functional working program as well as an educational program in entrepreneurship where scientists and students work closely with faculty from the school of business who can make valuable contributions to entrepreneurship that help to transfer discovery science into the clinic.

Entrepreneurship in the Education Mission

Traditionally, medical education has focused primarily on the transference of large amounts of basic science and clinical information to medical students, with success measured by students' achievement on national board exams. For the first 2 years, this transference takes place primarily through traditional lectures and didactic sessions. This environment, though rigorous, is typically one where many medical students flourish, because prior exhibition of academic excellence is usually one of the criteria for acceptance into medical school. But later in their third and fourth years, medical students encounter a different learning environment as they enter the clinical setting. It is here where students must transform from a solitary book-learner into a functional member of the clinical team. This transition may present a difficult challenge for some students, many having never before focused on the interpersonal and communication skills necessary to excel in this environment. Certainly, much of what comprises an exceptional physician are "soft skills," those skills that make them effective communicators, problem solvers, and team members, all skills at the core of entrepreneurial education. This type of education continues on into their years of residency training as well.

As mentioned previously, the past decade has seen a dramatic increase in formalized curricula for collegiate students interested in entrepreneurship. Over 90% of universities now offer courses or majors in entrepreneurship, many of which house dedicated centers of entrepreneurship within their schools of business. While the traditional pedagogic components of lectures and case studies are still prevalent throughout most curricula, there has been a recent push to focus more on experiential learning through games, simulation, and perhaps even active development of venture creation, often with a direct involvement with industry. These programs emphasize the importance of networking and team building, skills that are essential toward the creation of a successful entrepreneur.

In recent years, a few medical schools have recognized the value of integrating entrepreneurship teaching into their traditional curricula, tailoring the model toward the space of health care innovation. The Innovating Healthcare Solutions (IHS) program at the University of Texas Southwestern is one example of this integration. Inspired by courses in engineering design, this program immerses medical students into a program that takes them from problem to solution with a particular emphasis on problem identification, team building, and importantly, the engagement of community stakeholders.

While the potential for commercialization and revenue to the medical school and university remains an appealing benefit of promoting medical entrepreneurship during the education of undergraduates and medical students, it is important to recognize the value such experiences can provide for these students, some of whom will go on to pursue leadership roles in their respective fields. First, students pursuing entrepreneurship quickly learn the importance of interpersonal skills. The countless hours spent in meetings, competitions on the "pitch deck," and networking events teach students how to communicate effectively in a limited amount of time. Are such interactions much different than those experienced by the clinician during rounds or clinic office visits? Second, these educational activities/courses/experiences in entrepreneurship teach the value of collaboration across diverse disciplines. Often, successful medical start-ups or early stage businesses are composed of individuals from various areas of domain expertise. These interactions provide invaluable insights into how systems operate, preparing students who will soon enter an ecosystem that is both daunting and complex. Finally, entrepreneurial experiences teach students how to become problem solvers. They learn how to examine new environments through a fresh lens and become better equipped to identify inefficiencies and possible areas for improvement in health care delivery. Indeed, these young students often "see" solutions invisible to the long practicing physician or established researcher. In an age where the development of new communications and computing technologies far outpaces their integration into the complex system of health care, students with such experiences are well-positioned to lead the many new aspects of future health care delivery as these technologies make their way into the medical field. Therefore, it is in itself innovative for medical schools to integrate entrepreneurship into their educational mission.

Creating an Entrepreneurial Environment

An environment that encourages innovation in entrepreneurship begins with an engaged faculty that make it part of their daily activities to foster such an environment. Obviously, supportive leadership (and recognition) from the university president, board of trustees, dean, and associate deans is required so the faculty feel empowered to devote effort to entrepreneurship. In addition to leadership support, the necessary infrastructure should be present to fully develop a program in innovation and entrepreneurship. The infrastructure at a university should include knowledge experts in business development, legal advice, protection of intellectual property (IP), technology transfer, marketing, communications, and capital investment.

Entrepreneurship can also be encouraged through appropriate financial incentives. Typically, at a major research university, a well-defined policy addressing IP details the financial rewards that are returned to the university, medical school, and department as well as to the inventor. For example, at the University of Florida, an invention with a net adjusted income of $500,000 or greater would have the following distribution: 25% to the creator, 10% to the program, 10% to the department of the creator, 10% to the college of the creator, and 45% to the University. This type of distribution on an invention such as Gatorade (now the G series) returned several million dollars annually to the Department of Medicine. This type of financial return to a department allowed further investment in research activities thereby, enabling the possibility of establishing the virtuous cycle of innovation and entrepreneurship.

Establishing an environment of innovation and entrepreneurship is recognized increasingly as a necessary and positive program in a research university. The interest of faculty and students at research universities demand educational programs in this field, and thus, the university infrastructure should not only support the program but encourage engagement in innovation and entrepreneurship.

OFFICE OF TECHNOLOGY TRANSFER

The role of university technology transfer is described thoroughly in Chapter 4 by Schrankler. From a decanal perspective, the expectations of faculty members who create IP must be managed. Though all faculty members who have IP believe the university should license their potential invention, the university and office of technology transfer must make informed but fair decisions about the university's interest in the IP depending on the university's investment portfolio, the development costs, and the likelihood of financial return. Individual faculty members often are a bit naïve in this regard, do not have the full perspective of the role of an office of technology transfer, and may require counseling by the dean or department chair along with experts from the office itself. This can be a sensitive topic, but it is important for the innovator to realize that this office can help them to achieve their goal in ways that are mutually beneficial to both the university/medical school and the individual.

RELATIONSHIP BETWEEN THE UNIVERSITY AND COMMUNITY ENTREPRENEURSHIP

An innovation and entrepreneurship program at a research university can be enhanced markedly by a community that is invested in entrepreneurship. Numerous examples demonstrate the powerful alliance of a research university embedded in a community that hosts an innovation hub and/or incubator. In many of these environments, the community of innovation arose because of the availability of content expertise provided by university faculty. Some of the notable partnerships between a university and a community entrepreneurship business include Stanford University and Silicon Valley, University of North Carolina, Duke University, and North Carolina State and Research Triangle Park, and Washington University in St. Louis with Cortex and BioSTL. Many other examples of a vibrant, university-community partnership in entrepreneurship exist, not infrequently because of university investment in the community.

The benefits of a community invested in entrepreneurship are numerous. To highlight some of these advantages, we will use the example of Cortex and BioSTL in St. Louis, MO. BioSTL was the vision of Bill Danforth, the former chancellor of Washington University in St. Louis. With investments from Washington University, the Danforth Foundation, and the McDonnell Douglas Corporation, BioSTL was created to foster entrepreneurship in an environment rich in content expertise from Washington University and Saint Louis University, innovation infrastructure, and venture capitalists. BioSTL is the innovation and entrepreneurship program housed within the physical structure of Cortex, a 200-acre area in midtown St. Louis. Cortex was funded jointly by BJC HealthCare, Washington University, Saint Louis University, the University of Missouri-St. Louis, and the Missouri Botanical Garden. Since its inception in 2002, Cortex has grown to occupy 1.7 million square feet of space and has attracted over $550 million in investment and over 4200 technology-related jobs. Cortex houses BioSTL and other programs such as the Biogenerator, which provides laboratory space to start-up companies. The activities at Cortex and within BioSTL are stimulating and successful with numerous early phase start-up companies working diligently to take an idea from concept to commercialization.

Programs such as BioSTL and Cortex engage the entire community by having coalition members that represent several universities, the civic community, business leaders, and investment bankers. Numerous educational programs are available for interested parties. In addition, BioSTL has subsidiary organizations such as Global STL which aims to provide a direct connection between a global start-up company and St. Louis University or a St. Louis Company. The ultimate goal for GlobalSTL is for these global companies to establish a physical presence in the St. Louis community. To foster engagement between global startups and local scientists or leaders, GlobalSTL hosts day-long engagements in which the global company presents their wares to interested St. Louisans, and this event is followed by a reception rich in fellowship. The results of such a networking event include the use of new software from the company 3D4 Medical to teach anatomy to medical students. As another example, Zebra Medical, a company from Israel, has expertise in combining diagnostic imaging data, pathologic data, and potentially genomic data to improve diagnostic and prognostic accuracy that may be of value to clinicians at Washington University Physicians and St. Louis University.

The collaborations and relationships that develop from true partnerships between universities and community entrepreneurship programs are invaluable. Both parties bring valuable resources that can be shared to advance ideas to commercialization. Such programs are invaluable and the envy of all universities that do not have such opportunities.

PHILANTHROPY

One must not overlook philanthropic gifts from grateful students who gained their expertise in business entrepreneurship from their university or medical school. Not infrequently, these generous entrepreneurs wish to recognize the role of their outstanding undergraduate or graduate education in the development of a large, profitable company. They may choose to give a generous donation to the university or medical school in recognition of the thought-provoking education they received. These gifts are often used to support programs in entrepreneurship and, therefore, advance the cycle of innovation and entrepreneurship.

IMPORTANT POINTS

- Medical schools and universities are environments rich in the knowledge environment, and innovation and entrepreneurship should be fostered to take advantage of this knowledge and discovery.
- The formation of programs focused not solely on discovery but on conversion of these discoveries (concepts) to commercialization of useful products will enhance the university and medical school as well as the surrounding community.
- Increasingly, medical school deans will focus on the development of these programs that enrich the clinical, research, and educational missions of the school and serve to promote better care of the patient.
- The schools, faculty, students, staff, and communities will be rewarded for these efforts.

Chapter 23

The Role of Medical Societies and Their Foundations in Supporting Entrepreneurs: A View From Anesthesiology

James C. Eisenach
Foundation for Anesthesia Education and Research, Schaumburg, IL, United States

Medical societies exist primarily to serve their members. Although they may provide education and resources to the public and others, a majority of the efforts of medical societies focus on supporting the needs of their members. Medical societies also establish or support peripheral organizations of relevance to their members, ranging from political advocacy activities to foundations which support education, research, or history in the specialty. The Foundation for Anesthesia Education and Research (FAER) is such a foundation of the American Society of Anesthesiologists.

The mission of the FAER is to develop the next generation of physician-scientists in anesthesiology. Our programs aim to expose and recruit medical students and anesthesia residents to both the major public health problems in the specialty and to the anesthesiologist investigators who are addressing these problems; a second aim is to fund fellows and junior faculty beginning a research career. Our goals are to foster interest among medical students and residents in a research career and to help to launch successful research careers among young investigators. Thus, we center on individuals at the beginning of their medical career and on a mentored career development plan.

For its first three decades, the FAER had no programs aimed specifically at entrepreneurs, although in its first decade, the FAER supported very small "starter grants" that were not focused primarily on career development; a handful of these starter grants happened to have an emphasis or component focused on entrepreneurship. Beginning in 2013, a group of the FAER Board members and advisors led by Ted Stanley, MD, advocated for the relevance of entrepreneurial innovation to the mission of FAER; this concept generated

Medical Innovation. https://doi.org/10.1016/B978-0-12-814926-3.00023-1

discussions both on whether the FAER should pursue this line of funding and if so, how to do so.

All medical specialties include physician-led entrepreneurial innovators, but how medical societies and their foundations support them varies considerably, in part reflecting their mission statements. While we recognize that entrepreneurial efforts, just like science, serve importantly to advance medical practice, how and if the FAER can reconcile its mission to develop careers of physician-scientists with support of entrepreneurial activities is unclear. One of the goals of our most important donor, the American Society of Anesthesiologists which provides 50% of the FAER's income, is to develop through the FAER the scientific basis and advances of the specialty. In addition, our appeals to donors, both individuals and groups, emphasize career development of physician-scientists. Arguably, to apply these donations to specific entrepreneurial projects would betray the trust of these donors who have responded to solicitation of funds for career development or fundamental scientific advances. For these reasons, the FAER has avoided to date funding grants to entrepreneurs, because it was determined that such grants are outside the current mission of the FAER. The FAER does, however, recognize that entrepreneurship is a legitimate category of career development. Our Foundation also appreciates that technology development often creates substantial opportunity for future clinical and fundamental research, with pulse oximetry and ultrasonography being only two examples.

With these considerations, the FAER has experimented recently with programs which might meet some of the needs of physician entrepreneurs in the specialty, and several of these programs have been largely educational in nature. Two, separate, 1-day workshops were run since 2014. One workshop was a stand-alone meeting; the other workshop was conducted jointly and in collaboration with the annual meeting of the Society for Technology in Anesthesia (STA) and was branded as the Anesthesiology Conference on Innovation and Entrepreneurship. The focus of these workshops was to overview the natural history from concept to commercialization and to identify and address the road blocks encountered commonly and strategies to overcome these road blocks. Both workshops were well attended and reviewed by participants.

Subsequent discussions among the FAER Board members and advisors have concluded that the FAER should create a different type of approach with goals to increase the opportunities for networking between entrepreneurial innovators and those who fund and develop their ideas to the market. The format chosen for this was modeled on the television program Shark Tank. Featured entrepreneurs present their ideas to a panel of venture capitalists, other funders of early phase innovations and successful anesthesiologist inventors who chose the best applicants. All projects must have preexisting protection of their intellectual property, because the presentations are public. The sessions are promoted to both the potential pool of applicants and representatives of venture capital/industry with an aim to recruit a large audience that includes opportunities for networking between the two groups. At the time of this writing, we have completed the application phase of the first of these sessions, and the panel is evaluating which 4 of the 28 applicants will be chosen. The success of this program is evidenced by the number of individuals interested in presenting their ideas in this forum.

Philosophically, how the FAER in particular and the American Society of Anesthesiologists in general should address the needs of entrepreneurs in the specialty is unclear. The FAER, like many/most other medical societies, has historically funded career development grants to attractive/exciting projects submitted by junior faculty physician-scientists. We define and measure success by predefined milestones in career development during and after the award and adjust the program to increase the degree of success that we measure with these predefined milestones. The forays FAER has conducted to date in supporting entrepreneurial innovation have not defined a priori what a successful program looks like and how it should be measured qualitatively or quantitatively. That there is a need and interest in better educational offerings, networking opportunities, and support for physician entrepreneurs in anesthesiology is undisputable, but what role a foundation, with an explicit mission to develop careers of physician-scientists, has to meet these needs remains an enigma to the FAER as it does probably to many medical societies and their foundations. It is clear to the FAER, as it may be to many foundations, that education in innovation, and networking opportunities in its development are crucial for entrepreneurial innovation to flourish in medicine.

Chapter 24

Surgical Societies as Supporters of Innovation

Lee L. Sanström[*,†]
[*]*Institute for Image Guided Therapies, University of Strasbourg, Strasbourg, France, and*
[†]*The Oregon Clinic, Portland, OR, United States*

Other than serving as community forums for the latest developments in surgery, surgical societies have not been mandated typically to foster innovation as part of their mission. There are of course some exceptions. For decades, the exhibit hall of the American College of Surgeons (ACS) served as a shopping mall for surgeons to see and test the latest in surgical technologies from industry. As surgeons slowly lost the ability to influence the purchasing decisions of their hospitals, and as the ACS evolved to more of a political/social organization, the attraction of the exhibit hall has faded. Today, surgeons often look to subspecialty societies for their education and for an introduction to latest innovative developments.

THE BIRTH OF SAGES AS "THE SURGICAL SOCIETY FOR INNOVATION"

The Society of American Gastrointestinal and Endoscopic Surgeons (SAGES) is somewhat unique among the surgery subspecialty societies in its interest in facilitating innovation. This interest stems from its conception as a specialty focused on a technology, initially endoscopy, then laparoscopy, and now many forms of minimally invasive surgical procedures from the abdomen to the chest to the heart, and then to most other organ systems through a minimal access approach. The meteoric rise of SAGES in the pantheon of the house of surgery was primarily due to its proactive adoption and promotion of the "new" and "latest" advances in primarily medical devices but also to its leadership structure that has to some extent "institutionalized" adaptability to the "new."

Unique among surgical societies, with the exception of some very new societies focusing on robotic laparoscopic surgery, SAGES was created in response to technical innovation that allowed new technical interventions as opposed to the more traditional geographic or disease-oriented focus of most societies.

Medical Innovation. https://doi.org/10.1016/B978-0-12-814926-3.00024-3

As such, from its start, SAGES has always fostered a strong alliance with industry and has successfully leveraged innovations in surgery to grow in size and influence.

The founding of SAGES in 1981 was a direct result of the introduction of the flexible endoscope into medical practice in the late 1960s. Flexible endoscopy, as a revolutionary therapeutic innovation (device), was developed and promoted initially by surgeons, but gastroenterologists rapidly adopted this technology to study the lumen of the stomach and colon directly by using a fiberoptic camera incorporated into a flexible, drivable tube, the endoscope. By 1980, use of this technology was in danger of no longer being available to surgeons due to the "turf wars" promulgated by gastroenterologists [1]. To combat this danger, SAGES was founded to protect the rights of surgeons to practice flexible endoscopy and to promote this new technology through training, guidelines, and its annual meeting. SAGES found a ready ally in the flexible endoscopy industry who saw surgeons as a large potential market; indeed, initially, SAGES depended heavily on corporate sponsorship to help finance the society and to train the younger generation of surgeons in the use of endoscopy in the practice of surgery.

At the annual meeting in 1989, Jacques Perrisat, an innovative surgeon from Bordeaux, France, introduced the adoption of endoscopic techniques and technology to perform a "laparoscopic" cholecystectomy via small incisions to an amazed but endoscopic-minded audience [2]. This minimally invasive procedure to remove the gallbladder, as opposed to the conventional large abdominal incision, led to the further development and subsequent explosion of minimally invasive operations by surgeons; the demand to adopt this new approach was in many respects unprecedented in surgery. SAGES as a society wisely and proactively elected to take charge of the further development of this innovation by designing training courses, including resident training and "train the trainers" courses, by writing guidelines for the indications and use of MIS, and by becoming the leading forum for presentation and publication of early results and new technology (innovation) as the technology spread to other procedures and specialties. SAGES accomplished all of this very much in partnership with industry, both in principle and in the development and marketing of minimally invasive surgery. This working, mutually beneficial partnership started with the great innovators in the field, Ethicon, Inc. and US Surgical Corp., but also with a multitude of smaller medical instrument and engineering companies and many new startups spawned by this surgical revolution.

The benefits of this close relationship to industry and to SAGES (as well as to the patient!) were apparent immediately. SAGES grew from a membership of a few hundred to over 3000 in only a couple of years and to almost 7000 today—the largest general surgery society in North America. The annual meeting of SAGES became the go-to place to learn about the latest developments and, of course, the medical device companies were also benefited tremendously. In fact, industry benefited not only from product sales but also from new product ideas generated by the innovative minds of the "techies" who were drawn to SAGES. Moreover, at the annual meeting of SAGES, industry was provided an attractive venue to exhibit their technology and benefitted from being both visible and approachable to this large surgical society. SAGES was able to bring a level of control to the rapid and sometimes haphazard adoption of this new technology. SAGES in turn, made itself the home for this and related industries, forging a long lasting, mutually beneficial relationship, both financially and from the aspect of new, innovative, health care delivery based initially and primarily on new technology, that is, innovation. Unique for a surgery society, SAGES founded a Corporate Council in 1991 which was very active in its interaction with industry both for research grants to develop new ideas as well as for education of the next generation of surgeons. This Corporate Council had a direct line of communication with the leadership of SAGES and directly with several of the committees of SAGES. Each group proposed projects to the other, and industry was lavish initially with its funding. This fertile frenzy of innovation and development of the newest of techniques in minimally invasive surgical procedures persisted throughout the 1990s, guided (albeit somewhat intuitively) by SAGES and its members, and established SAGES as the surgical society of the new, the young, and the innovative surgeon of tomorrow, a role it has struggled, more or less successfully, to maintain today.

THE TIMES THEY ARE A CHANGIN'...

Moving out of the amazing decade of the 1990s when new surgical technologies were introduced monthly, new procedures were developed based on these technologies, and a whole new way of thinking of surgery (minimally invasively) was adopted worldwide, but things began to change. The FDA made the process of introduction of new medical innovations more difficult and expensive—and as a result, the pipeline of new devices dried up. This change was compounded by the fact that surgeons came to realize that they had been giving away their ideas (their intellectual property) for free to industry, and as a result, their interactions become more entrepreneurial, changing their relationship with both industry and society meetings. Another great suppressor of the free exchange of innovation with societies like SAGES was the rise of the concept of conflict of interest (COI) [3]. Strict COI regulations were imposed on medicine and surgery in particular. This process was partly in reaction to some of the excesses of the 1990s laparoscopic whirlwind and partly a self-serving move on the part of a rapidly consolidating industry. The 1990s resulted in multiple surgical device startups which began to be acquired by larger companies, and the rapid growth and profits of the major players made them targets for even larger multinational companies. These bigger public companies were more conservative and had less loyalty to their customers than the smaller entities. In addition, they were interested in escaping the now established tradition of heavy financial support to the surgery societies. The AdvaMed Code of Conduct [4] and other mandated

COI initiatives were therefore, met with little protest from industry and the exchanges between industry and surgical societies became more distant and less trusting. SAGES itself was changing as well. No longer the "outlaw" radical society, the more mature and certainly larger SAGES sought to distance itself from the free for all but innovative past and began concentrating on a higher level of evidence-based publications and presentations, more "serious" subjects at meetings, and on being a good citizen and avoiding appearances of COI. Although this dampened the response time and near instantaneous uptake of innovation, SAGES was still the place surgeons came to learn about the "new" as well as highlight their ideas.

Of course, innovation in surgery is a fairly constant phenomenon—however, everyone continues to look for the next "disruptive" innovation like laparoscopic surgery turned out to be. So far, nothing has risen to that level. Single-port laparoscopy generated mild enthusiasm and was accorded special sessions at the annual meeting, but faded fairly rapidly from the radar. Natural Orifice TransEndoscopic Surgery (NOTES) occasioned a greater response from surgeons and SAGES (as well as the other surgery societies). It is interesting to note that SAGES, in its more serious and grown-up mode, chose to explore the NOTES innovation by cofounding an outside entity with the American Society of Gastrointestinal Endoscopy (ASGE) [5]. After a few years of excitement on the part of surgeons and industry, NOTES was by and large slowly abandoned by the societies after a high-volume procedure was not discovered. Robotic surgery is an example that is rapidly gaining traction and may end up being one of these disruptive innovations, albeit an extension of laparoscopy, but further development of the technology is needed and it is a positive observation that competition in the robotic industry is starting.

SAGES AND INNOVATION TODAY

In spite of being a more serious, thoughtful, and conservative society today—consistent with its size and increasing role in American surgery, I should add that SAGES is still regarded as the best venue for presenting or publishing innovation. The following is a list of some of the methods SAGES uses to fulfill its mission of promoting innovation:

Committee structure: SAGES continues its focus on young surgeons via its unique committee structure. Committees and committee-like initiatives (e.g., task forces) are well populated, and usually chaired by young SAGES members. There are currently 35 listed committees and task forces—meaning that hundreds of SAGES members (currently more than 700) have a direct say in the workings and future directions of their organization, a phenomenon unlike most other societies. There are currently two committees directly responsible for innovation: the Technology Assessment and Valuation Committee (TAVAC) and the i3 Summit task force (innovators, investors, and industry). The TAVAC committee evaluates new technologies that are already FDA-approved and on the market [6]. The committee performs a formal and very exhaustive review of the literature and then publishes their result in the journal SURGICAL ENDOSCOPY—generally concluding that as of yet, there is insufficient high-level evidence to make a conclusion. While thorough and evidence-based, this is somewhat frustrating to both SAGES members and industry, who often would like early indications of what technologies are interesting or need a strong endorsement by a respected surgical society to make it to market. An example is the recently published TAVAC review of the Linx antireflux device (Torax, Shoreview, MN) which appeared online on August 2017, some 5 years after Linx had been introduced clinically [7].

In response to criticisms of the TAVAC as not providing the needed support of innovation, SAGES last year (2016) created the i3 Summit task force. The mission of this group is to re-establish that direct link between SAGES the society, its innovator members, and industry that existed in the 1990s. The exact mechanisms of this relationship remain to be fully worked out, but it is a promising move to re-establish SAGES as a leader in vetting new technologies and techniques.

Annual meeting: The annual meeting of SAGES continues to attract around 2000 attendees each year. Aside from serving as an important social meeting point for inventors, clinicians, and industry, the meeting has several areas of interest to innovators. The exhibit hall remains a lively source of seeing and on occasion even experiencing the latest technology; this venue is used widely by industry to interact with key opinion leaders, called the KOLs. For the past several years, SAGES has highlighted the "Emerging technology" session, which specifically targets innovations and techniques that have not yet reached "prime time," that is, ones are still in development or have only limited clinical data. This session has proven to be one of the most popular sessions at the annual meeting. An added benefit is that the abstract submission deadline is more than 2 months later than the deadline for submission for the main sessions [8]. These sessions are typically "standing room only" and generate more discussion and "buzz" than almost any of the others. The program committee, perhaps mindful of the history of laparoscopic cholecystectomy and SAGES, also keeps a slot open in the program (generally designed a year or more in advance) just for any late arriving innovations. An example of this is a last minute proposal for the 2018 program to include a session on interventional radiology for surgeons as a new horizon in minimally invasive surgery [9]. Although proposed by Dr. Eran Schlomovitz 6 months after the program had been created, it was considered innovative and important enough to wedge it into the session line-up (personal communication).

Foundation and research grants: Like most surgical societies, SAGES has a mandate to foster and fund research, particularly for young surgeons interested in an academic career. While there are four types of grants available, the ones potentially able to foster innovation are in the category of research grants. These $30,000 grants are judged based on their scientific quality, but innovation and a focus on new therapies always receive extra merit. Sadly, in spite of its healthy Foundation, the number of these research grants has decreased over the last decade due to decreasing support from industry. Interestingly, last year Intuitive Surgical, amazingly enough for a company notorious for not supporting research or education, funded two separate $50,000 grants to study robotic surgery [10]. Of course, these were limited to "commercially available" surgical robots (i.e., Intuitive Robots) and thus, they are not really geared toward innovation in the field of surgical robotics.

CONCLUSIONS

Surgical societies can certainly play a role in surgical innovation. Their meetings provide a forum for industry and thought leaders to meet in the exhibit halls and as well as on the actual program, although their importance, diluted by internet access and loss of physician input into technology purchases at the hospital level, remain popular and educational. SAGES in particular, having been catapulted into prominence by its rapid uptake of two disruptive innovations in surgery, remains sensitive to the need to maintain its "finger on the pulse" of new advances. In spite of its increasingly serious role in American surgery, SAGES is proud of the fact that it is still regarded as the "home of innovation" in surgery and therefore, must remain open to and vigilant for the next disruptive innovation. To date, in spite of some minor ups and downs, it seems to have managed this balance well.

IMPORTANT POINTS

- SAGES is a society that focuses on a technology and not on a disease process or organ system, allowing it to foster growth and development in innovation.
- SAGES has continued to foster interaction with industry to further the development of MIS techniques and other new devices.
- Recent challenges to this interaction have included the introduction of concepts such as protection of intellectual property, conflict of interest, and the increasing university and federal regulations for industry partnerships.
- SAGES offers several grants designed specifically to promote innovation.
- Surgical societies can certainly play a role in surgical innovation. Their meetings provide a forum for industry and thought leaders to meet in the exhibit halls and as well as on the actual program.

REFERENCES

[1] Zetka JR. Surgeons and the scope. Ithaca: Cornell University Publishing, ILR Press; 2013.
[2] https://www.sages.org/video/the-history-of-sages/.
[3] https://www.sages.org/accme/policy-commercial-support-conflict-interest/.

[4] https://www.advamed.org/.
[5] Hawes RH. Transition from laboratory to clinical practice in NOTES: role of NOSCAR. Gastrointest Endosc Clin N Am 2008;18(2):333–41.
[6] Krause CM, Oleynikov D. Evolving responsibility for SAGES-TAVAC. In: Stain SC, Pryor AD, Shadduck PP, editors. The SAGES manual ethics of surgical innovation. AG Switzerland: Springer Publishing; 2016. p. 223–8. May 5.
[7] Telem DA, Wright AS, Shah PC, Hutter MM. SAGES technology and value assessment committee (TAVAC) safety and effectiveness analysis: LINX® reflux management system. Surg Endosc 2017;(August 25). https://doi.org/10.1007/s00464-017-5813-5.
[8] https://www.sages.org/2018-call-abstracts.
[9] https://www.sages2018.org/program/.
[10] https://www.sages2018.org/grants.

Chapter 25

Role of Medical Journals in Promoting Innovation

Michael G. Sarr
Mayo Clinic, Rochester, MN, United States

One might think that a journal editor is ideally suited to promote innovation and technology. You should be correct, however, there are many other considerations, some parochial for the journal, others related to the reputation of the editor and the publisher, and finally the concept that many potential "innovations" and "innovative ideas" are often not yet proven. Why is this so? Serious scientific journals go out of their way to publish *only* evidence-based studies and are not comfortable in the space between innovation and clinically impactful data, especially concerning new devices, first-generation medications, and novel treatments. This approach is in direct contrast to many newspapers, magazines, and periodicals which troll for something dramatic that will challenge current medical treatment, promise cures, and offer hope for those desperate people just to increase their circulation, even if there is no or minimal evidence that it works. The understanding is that high-impact scientific journals live by a code that *evidence-based* science is supported with depth and substance, and addresses an unresolved research question that has been peer-reviewed by experts in the field. Yes, the topic maybe "disruptive," which can be good and may very well lead to changes in the field of medicine or science that leads to a major paradigm change in thought or practice, but there is no obligation or "code" that scientific journals should promote "theoretically attractive or seemingly/potentially important ideas of innovation" that support a new concept, diagnostic test, therapy, or "device" that is purported to be a new important innovation.

First, let us start with a working definition of innovation or at least, the definition that will serve as the basis of this editor's outlook on the role of medical journals in this arena of innovation. Innovations are either proven to be of worth or are still in the development stage of theoretically attractive ideas that seem to make sense but are, as of yet, not proven. There are "innovations" that are of proven value that have been tested in a statistically valid study to show benefit—these are the easy ones that editors strive to attract to their journal. While the bar for acceptance for publication may be a bit lower than for other

Medical Innovation. https://doi.org/10.1016/B978-0-12-814926-3.00025-5

articles, there still needs to be credible, scientifically valid data supporting the claims of "innovation" by the author(s); these type of articles may be pilot studies that provide the basic evidence that the idea has merit, but they may not have been designed in such a way to provide incontrovertible data or statistically valid results. In contrast, there are other "potentially innovative ideas" that are as yet unproven but offer the possibility of being disruptive, that is, treatment-changing breakthroughs. For these nonproven innovations though, the idea behind the innovation is attractive, seems reasonable, and is often exciting, but without solid evidence of efficacy, most all respectable, top-shelf journals will hesitate to publish these "discovery-type" articles for fear of repercussions for the journal's reputation and the fallout of the claims of the author(s) if they prove not to be true.

The mantra of the editor of a serious scientific journal is to promote progress in science and even discovery, but when it comes to claims of new treatments, new devices, or new medications, convincing data are paramount. This understanding serves as the principle of the time-honored, peer-review process. Submissions are reviewed by experts in the field to determine whether the study has fulfilled the goals outlined in the introduction and whether the submission supports or fails to support the hypothesis offered by the investigator. Conclusions unsupported by data or "opinions" from "self-appointed experts" (e.g., the authors) without objective evidence to support these claims or opinions are not acceptable, and though often picked up and exaggerated by the lay press, serious medical journals do not publish unsolicited opinions or hypotheses without results to support the claims.

There are various layers to this editorial stance. For instance, Phase I clinical trials of medications or treatments are designed to test the safety, *but not the efficacy,* of the new treatment. These new treatments have some evidence of potential efficacy based either on known responses to similar treatments or some preliminary laboratory data supporting the mechanism of action. When published, the goal is not to study efficacy but rather to use dose-escalation to determine the optimal dose of the medication with the fewest side effects; these studies are especially common with cancer trials, where new drugs found to be effective in cell culture or animal models of cancer are tried experimentally in man. Everyone should know that a Phase I trial is not a definitive statement on efficacy, even if the outcomes in these smaller, nonstatistically valid studies show dramatic effects. Even the next step in some treatments, the Phase II trials, are designed to recruit more patients and to look for findings highly suggestive of efficacy prior to undertaking the definitive Phase III trial designed to perform a robust, adequately powered study based on expected results and with the bar set high enough to allow the critical reader to not only make a statistically valid conclusion but also a conclusion that is clinically relevant.

So, what is the responsibility of the journal editor in promoting innovation, one of the current hot topics in medical journals? The editor is the last "reviewer" to decide whether the innovation is worthy of publication based on peer review, as well as the editor's review of solid experimental evidence. Premature publication before statistically valid efficacy has been demonstrated can be potentially harmful to patients or the medical system. All editors crave the visibility of being the "first" journal to publish new, innovative treatments, especially the disruptive, game-changing ones that, when published, will attract readers and importantly citations from other publications (these latter citations serve to increase the impact factor, which in some publishers' and editors' minds serves as the currency of the journal). But these same editors fear the very real fallout if the premature publication of a new, purportedly effective treatment is found later to be ineffective, dangerous, related to unsound scientific studies, or composed of premature, unfounded suggestions of efficacy reflecting the inadequate peer-review process of the journal; this unfortunate situation, though rare, leads to bad press for the journal (and for the editor) and discourages other authors from publishing their best work in that journal. Thus, editors must promote and protect the reputation of their journal (and their personal reputation) and, therefore, will often not risk supporting a potentially risky "innovation" that may prove later to be ineffective or even harmful, *even if* the journal publishes a disclaimer by the editor(s) about the article saying that the article is highly suggestive of a new development, but the data do not confirm the authors claim. It should be noted though that some editors will bend these principles on occasion, and it is the opinion of this author that such a policy may be necessary in today's world of innovation.

All this does not necessarily mean that a journal cannot promote certain aspects of innovation. I am going to use my journal SURGERY as an example (I have been an editor of SURGERY for 20 years). SURGERY has developed recently several sections devoted to some of the principles of innovation. First, innovation is not only new medications, cell-based therapies, or new devices allowing more extensive interventions in a safer way; innovation may also be in the novel ideas used to deliver heath care. For instance, we published what we called low-tech innovation in the developing world. The group Helps International is a nonprofit, humanitarian agency that delivers both basic health care as well as providing several other means of personal health and safety for the indigenous people of Guatemala [1]. The leadership of Helps International noted that there were an inordinate number of serious burns in the children of these indigenous people. Also, many of the parents had a chronic cough and had reactive airways. Further study of their home environment showed that these people cooked over an open, wood burning fire on the floor of their homes; there was no chimney and the particulate smoke filled the room and exited the home via the eaves of the roof. An "innovative" member of Helps International, Mr. James O'Neal, proposed making an inexpensive "stove" out of a type of cinder block structure containing an inlaid firebox; this "Onil Stove" served several innovative purposes: (1) it got the fire off the floor, protecting the young children from burns, (2) being a more closed system, the smoke was captured in a chimney and directed out of the house without polluting the air inside the house, and (3) the firebox was 70% more efficient than an open fire, thereby, helping not only to prevent deforestation by the constant need for firewood, but also relieving the women from the daily, time-consuming chore of finding enough firewood to maintain the open fire in their house—this freed the women to pursue other activities/opportunities. This low-tech "innovative idea" has resulted in over 250,000 of these stoves having been built on site and installed in the homes by the Helps organization over the last 28 years. In addition, a simple, effective, inexpensive water filtration system, which has decreased much of the childhood diarrheal diseases, was developed later and is now also being established in these homes. The impact on the health and well-being of these indigenous peoples is far reaching and was recognized by the prestigious Ashden Award for innovation [2].

Similarly, in Tanzania, a different group has worked with a local hospital to try to develop means to raise money locally to provide continual, onsite financial support for the hospital built by this humanitarian group called the Dodama Tanzania Health Development based in Minneapolis [3]. Innovative (imaginative) approaches have included an onsite inexpensive system to build long-lasting home building blocks from the local soil and to sell them for a small profit [4]. Similarly, by finding a clean source of potable water via drilling a well has allowed the hospital group to bottle the water and sell it to the community, thereby decreasing the incidence of childhood diarrheal diseases. Again these simple, imaginative innovations are important. These differ a bit from the principles of the majority of this book, but nevertheless, they offer unique innovative possibilities to help societies. There are many similar possibilities for innovation in the developing world, many of which may be profitable.

In this same section of innovation, we published a report on a technique to de-cellularize the liver while maintaining the extracellular matrix. In contrast to the low-tech innovations described earlier, this "high-tech," patented, proprietary process, while as yet unproven, is theoretically attractive in that it deletes the liver of cells which contain the major histocompatibility (MHC) Class 1 and 2 antigens involved in immune rejection, yet preserving intact the native, three-dimensional architecture of the liver, which "in theory" might allow the seeding of the extracellular matrix with cells of a different genetic makeup as a form of regenerative medicine. While, at the time of publication, this company had yet to prove the efficacy of this regenerated liver in transplantation, the concept was attractive. We were also very careful, however, not to claim true efficacy, but rather, to promote further experimentation using this innovative technique [5].

We also have a section called "Hypothesis," designed to allow authors to suggest an innovative idea that is as yet unproven but has considerable experimental data supporting the hypothesis. Currently, it is difficult to "get out" an attractive, but unproven hypothesis, of a serious investigator without any studies to either prove or disprove the hypothesis. Again, we are careful to select the hypotheses that seem reasonable but that are also based on prior objective, experimental work relevant to the topic and to our readership. This interest in publishing innovation articles represents a small fraction of our publications, but these types of academic and scientific reports of interest are consistent with our mission to report new ideas relevant to surgical science.

Another example from our journal highlights the fine line between innovation and myth. We were approached by an author who proposed a new technique that would support the restoration of electrical continuity across the site of a complete spinal cord transection under the ostensibly outlandish topic of "head transplantation," that is, transplanting a deceased donor body onto the preserved head of a patient with a progressive, degenerative, peripheral neuromuscular disorder that will lead relentlessly to paralysis and death. While this seemed ostensibly like science fiction, with a detailed and intense review of the literature supporting an innovative technique that allows sharply transected nerves to re-fuse their membranes without the Wallerian degeneration we all learned in medical school, we reviewed the submission. This topic was discussed with the editorial board of our journal, several of whom critically reviewed the submission, and the discussion among our board was lively; several members felt this topic was too outlandish, while others believed that there was enough scientific data to support further study. We published a miniseries of three reviews of the early results of forays into this topic [6–8] not because of an interest in head transplantation (for which there are far too many ethical issues at this time), but rather as a possible technique for restoring electrical continuity (and thus movement and sensation) across the site of a previous, paralyzing, spinal cord injury in patients with a distant traumatic spinal cord transection. This was a bit of a risk, because several neurologic and neurosurgical journals had refused to publish this work. *So, here is where a journal and hopefully an open-minded editor(s) and editorial board can help to foster the visibility of an innovative idea*, the real topic of this chapter. This foresight and potential insight by our journal based on similar scientific studies in mice [9] and rats [10] eventuated in a proof of principle in a dog model, where the thoracic spinal cord was completely transected, the site of transection was treated with a neuroprotective solution, and objective electrophysiologic measurements of re-innervation across the site of spinal cord transection began to be apparent within 2 weeks of the procedure—something that no one based on our classic teaching in nerve transection/regeneration would have believed possible [11].

As an editor, I maintain that many journals are too concerned about their impact factor and of being accused of publishing science fiction or poorly researched work or opinion. Some journals may be too afraid to take a chance, or what may be called a calculated risk based on minimal scientific evidence. There are journals such as biomedical engineering publications that predominately address medical device research and deal with innovative designs from the aspect of mechanics and mechanical theory, but in the world of health care delivery. High impact, clinical journals are possibly too risk averse in their process of mandating a complete, statistically valid story/study before they will publish an innovative idea that has potential merit. Yes, there are some journals that do acknowledge that there are ideas, procedures, and devices which challenge current thinking and current dogma that deserve publication, *BUT* the innovation must be based on objective evidence highly suggestive of efficacy.

In summary, the role of top shelf, scientific medical journals is to publish strictly valid, evidence-based studies. This is the current time-honored dogma. But in a world of constant innovation, perhaps there should be a role for publishing in some form the less well-validated, new innovative ideas that deserve visibility and have potential merit.

IMPORTANT POINTS

- Currently, there are few top shelf journals that will publish potentially innovative ideas, devices, or therapies without strong, scientifically valid evidence supporting the proposed innovation.
- The reasons for this reticence/fear of publishing unproven, potential innovations, even if they are theoretically attractive and have a modicum of experimental support is understandable—serious scientific journals fear the repercussions of accusations of an inadequate peer-review process or the criticism of publishing the sensational article claiming a true breakthrough which later proves to be poor science or unfounded in fact.
- Should journals have a section to attract and promote new ideas, devices, or therapies, many of which may be "disruptive," despite the lack of solid, statistically valid, evidence-based studies supporting the concept? This author thinks yes, but under carefully controlled criteria with at least some objective evidence supporting the potential innovative idea.

REFERENCES

[1] Schultz P, Boeke G, Miller S, O'Neal D. Onil Stove: innovation in the highlands of Guatemala. Surgery 2017;162(1):9–11. https://doi.org/10.1016/j.surg.2016.12.002.

[2] www.ashden.org/ashden_awardswww.ashedem.org/winners/helps.

[3] www.dthd.org.

[4] Gamble W, Toso J, Griffin B, Griffin B. Development of a sustainable project in Tanzania based on simple, innovative ideas. Surgery 2017;162(1):7–8. https://doi.org/10.1016/j.surg.2016.11.032.

[5] Seetapun D, Ross JJ. Eliminating the organ transplant waiting list: the future with perfusion decellularized organs. Surgery 2017;161:1474–8. https://doi.org/10.1016/j.surg.2016.09.041.

[6] Canavers S, Ren X, Kim CY, Rosati E. Neurologic foundations of spinal cord fusion (GEMINI). Surgery 2016;160:11–9.

[7] Ren X, Orlova EV, Maevsky EI, Bonicalzi V, Canavero S. Brain protection during cephalosomatic anastomosis. Surgery 2016;160(1):5–10. https://doi.org/10.1016/j.surg.2016.01.026.

[8] Ye Y, Kim CY, Miao Q, Ren X. Fusogen-assisted rapid reconstitution of anatomophysiologic continuity of the transected spinal cord. Surgery 2016;160:20–5. https://doi.org/10.1016/j.surg.2016.03.023.

[9] Kim CY, Oh H, Hwang IK, Hong KS. GEMINI: initial behavioral results after full severance of the cervical spinal cord in mice. Surg Neurol Int 2016;7(Suppl. 24):S629–31.

[10] Ren S, Liu ZH, Wu Q, et al. Polyethylene glycol-induced motor recovery after total spinal transection in rats. CNS Neurosci Ther 2017;(June 14). https://doi.org/10.1111/cns.1271.

[11] Liu Z, Ren S, Fu K, Wu Q, Wu JH, Hou L, Pan H, Sun L, Zhang J, Wang B, Miao Q, Sun G, Bonicalzi V, Canavero S, Ren X. Restoration of motor function after operative reconstruction of the acutely transected spinal cord in the canine model. Surgery [in press].

Chapter 26

Inspiration, Perspiration, and Perseverance: An Innovator's Perspective☆

Josh Makower[*,†,‡,§] **and Lyn Denend**[*,†]
[*] *Stanford School of Medicine, Stanford, CA, United States,* [†] *Stanford Byers Center for Biodesign, Stanford, CA, United States,* [‡] *ExploraMed Development LLC, Mountain View, CA, United States,* [§] *New Enterprise Associates, Chevy Chase, MD, United States*

What drives a person to innovate? [1] Fundamentally, I believe it is a profound desire to make a difference that compels most individuals to pursue the path of an innovator. And what better field is there to try to make a difference than in medicine, where new solutions are needed desperately, and their impact can be truly life changing (if not lifesaving).

Over the course of my career as a physician, inventor, and investor, I have observed that physicians are natural innovators, because they are faced constantly with unmet needs and problems they are eager to solve. Innovation is a natural calling to many in the medical field because of our innate desire to improve lives. As practicing physicians, we are able to make this positive impact one patient at the time. But for many, this is not enough. We have the craving to help more people on a substantially larger scale by driving changes in the way care is delivered, making improvements to the tools at our disposal, or even creating interventions that never before existed to help patients.

You have probably had ideas for new solutions at some point in your career—products, treatments, or services that could make a positive difference in the lives of many. Have you acted on them? Did you file a patent, build a prototype, or consider starting a company? If not, why? For many people, the prospect seems incredibly risky, intimidating, or overwhelming. For others, innovation or entrepreneurship becomes a calling in itself. Often called "serial entrepreneurs," these individuals may launch three, four, or even more

☆Although this chapter is written by both Dr. Josh Makower and his colleague Lyn Denend, selected parts of the chapter are written in the first person to connote the senior author's personal stories and viewpoints.

Medical Innovation. https://doi.org/10.1016/B978-0-12-814926-3.00026-7

companies during their careers. What separates those who develop new technologies from those who do not? Passion may be one factor; tolerance for risk another. But all too often, people are deterred because they do not believe they have the information, expertise, or resources to make their vision a reality. It is for this group of individuals that we were compelled to write this chapter and that we, with our colleagues at the Byers Center for Biodesign at Stanford University, created our textbook on the process of need-driven innovation in medical technology [2].

After devoting my career in large part to innovation in medical technology and spending years training other aspiring innovators as part of the Stanford Biodesign team, I am convinced that with the right amount of determination and perseverance, anyone can become an innovator. What they need is not an "aha!" moment of insight or inspiration, but rather a rigorous, proven process to use as a roadmap to get from identifying and understanding an important unmet medical need, to inventing a compelling new solution, and to implementing this potential solution into patient care. Describing that process in detail is beyond the scope of this chapter, but we would like to share a few highlights of the Stanford Biodesign approach, along with some of the lessons I have gleaned in founding eight medical device and health technology companies over the past two decades. Our hope is that this information encourages nascent innovators to rise to the occasion and play a more scalable role in helping patients and improving health care.

NEED-DRIVEN INNOVATION

So what do we really mean by need-driven innovation? In other scientific fields, an interesting discovery is often made first, and only then is there a search to find a problem it can solve. In consumer technology, at times the innovation cycle starts with something cool or attractive, and the product itself is used to capture the imagination of customers who do not even know they want it yet. This path could not be more different for medical technologies. The costs, time, and barriers associated with bringing a new medical device to market preclude the development of "technology for technology's sake." As a result, gaining a deep understanding of the need and its true value is vital at the outset of the innovation process. Understanding everything there is to know about a problem, knowing every angle of the condition, and examining each aspect of a clinical or health need—including the human impact, cost to the health care system, gaps in the existing landscape of treatment, and drivers of behavior and adoption within that landscape—all become critical elements of the inspiration for the right solution. As an entrepreneur, you must view your solution as the missing piece of a puzzle. To find the missing piece that will fit perfectly into the gap, you must first assemble all of the other pieces to get the clearest possible picture and the most complete understanding of the problem. Need-driven innovation starts with developing this in-depth understanding of the unsolved problem. Then, you can use all of the knowledge you have gleaned to guide a smarter, more focused approach to ideation and the development of a solution.

My start in medical technology innovation began at Pfizer in its medical devices business. There, the CFO/Vice President of Strategy asked my group to investigate why start-up organizations seemed to become less innovative after being acquired by a larger company like Pfizer. To accomplish this, I interviewed many of the founders and management teams of all the medical device companies that had been purchased by Pfizer to study their processes for new product development from the very first idea on which they founded the company all the way through their postacquisition activities. What I learned in many cases was that, when the innovators were first looking for what to work on, they started with a problem—or what we now call an unmet need—and then devoted themselves to developing an optimal technology to address it. But once they had a technology, their attention tended to shift to doing whatever was necessary to improve their core product incrementally or to advance it into new or adjacent areas. These start-ups tended to become laser companies or balloon angioplasty companies with their technologies viewed as their core competence rather than retaining the open-ended, need-driven process that they used when starting their organizations. Maintaining a deep understanding of the other needs of their customers and being open to venturing into completely new frameworks in technology to solve those problems, for the most part, became a skill set of their past instead of their future. As a result, much of their innovation and early growth stagnated.

That realization motivated me to begin developing the need-driven approach to innovation that is now the heart of the Biodesign innovation process. I had the chance to test and refine this approach on multiple projects over several years at Pfizer, but ultimately, I decided to become an entrepreneur to be able to apply it without the many constraints that exist within the large company environment. We raised the preliminary funding for ExploraMed Development, LLC, a medical device incubator that we founded in 1995, by convincing investors that need-driven innovation in the medical technology domain is a superior approach to the traditional technology-push model. Then, we set out to prove it. Years later, having launched multiple companies using this approach, I can say honestly that the process works and that it is an effective guide for first-time and experienced innovators alike as they seek to make a difference in the lives of patients through medical technology.

PICKING WINNERS

Compelling needs come in all shapes and sizes, so it is not always easy to tell which ones are worthy of your time and energy. Within the Biodesign process of innovation, our approach is to start by identifying LOTS of needs. Ideally, this outcome occurs by directly observing the care being delivered and watching the behaviors of affected stakeholders, including patients, family members, physicians, and other care providers. You can read about problems and opportunities in the literature, but fundamentally there is no substitute for observing them directly and bringing your own fresh, unbiased views to the challenges these stakeholders face every day.

Then, through an iterative, research-filter-research cycle, we recommend making the needs you have observed objectively "compete" against one another by comparing them on important factors, such as the size of the patient population, the extent to which a new solution could potentially change patients' lives, the availability (or lack) of existing solutions, and/or the degree of physician dissatisfaction with available technologies. Through this impartial exercise, it is possible to create a prioritization of the needs that seems to represent the most compelling opportunities.

Another important activity is to talk with peers, including clinicians, administrators, and even purchasers, about the needs under consideration and the value that a new solution can potentially deliver. The purpose of these interactions is to validate your understanding of the problem, how it impacts them, and how motivated they are to consider new alternatives. More often than not, compelling needs are linked to clear pain points of the patient or physician and a hunger for "something new."

As an example, with the ExploraMed team, we decided we would do something important to help new mothers. We started exploring needs in this space by researching common challenges faced by moms in the early stages of their postbirth experience. Energized by our desire to make a positive difference, we started gravitating toward big, complicated challenges that would require "blue sky" thinking to address. While we had many of these big ideas bouncing around in our heads, we convened a meeting of mothers to talk with them in an unguided way about the needs they felt were most important to address. To our disappointment, they expressed almost no interest in any of the areas we thought would have been important, and they overwhelming directed us to needs related to the unpleasant, inconvenient, uncomfortable activity of breast pumping. We had come across this issue in our research, but in our ignorance and naivety had set it aside, because improving the traditional breast pump seemed like such an incremental undertaking and, on the surface, we did not know how much of an impact it would have. After hearing the impassioned pleas of mothers for a better solution, however, we decided to take it on. When we opened our minds, what we heard loud and clear was that these mothers wanted mobility, discretion, a quiet pump, and good performance—a combination of features that had never been offered before—and that if this could be achieved, it would be transformational in the lives of mothers everywhere. That validation led us to create the Willow breast pump, a completely wearable, cordless, discrete, and quiet alternative that directly addresses the shortcomings of previously available technologies.

RUTHLESS OBJECTIVITY

After you become an expert in your field of need and then conceive of a potential solution that you believe will truly address it, the activities that follow should look more like destruction than creation. You must question your idea ruthlessly and objectively, tear out the parts that do not make sense, abandon, reinvent, and refine your concept over and over until it can withstand every challenge you can imagine. It is important to evaluate constantly and grapple continuously with every potential shortfall and issue associated with the idea, all the while trying not to "fall in love" with your solution. Infatuation with the one's own ideas before they are vetted properly has been the downfall of many innovators.

I learned this lesson through valuable firsthand experience. Several years ago, the ExploraMed team was working in the field of sleep apnea. After investigating the need thoroughly, we started generating various concepts for an improved solution. One of our favorite ideas was a sling for the tongue that would prevent it from falling back in the throat and blocking the airway. We designed a device, experimented with it on ourselves, and really thought we were on to something. The initial design would have to be more comfortable, so we set a goal of making it comparable on comfort yet more effective than the already available mandibular advancement devices (MADs)—a common solution for patients with sleep apnea. These MADs involve the use of a mouth piece to move the jaw forward to increase the diameter of the upper airway. We iterated the design multiple times and ran two clinical studies of our device trying to demonstrate superior results compared to MADs. Unfortunately, our sling required the patient to meter how much of their tongue they would willingly introduce into the device, which ended up being a huge variable that negatively affected our results. We were convinced that it would work better than MADs if the patient would use it correctly. But, in the end, after spending millions of dollars, reluctantly we were forced to abandon the idea as a result of this patient-centered barrier. The sad thing was that we had overlooked, or rather chose to ignore, evidence supporting the magnitude of this important patient-centered barrier much earlier in our design and development process, but our "love" for the idea and its simplicity blinded us to its shortcomings.

Painful episodes like this one illustrate that sometimes abandoning an idea you love is the only way to truly make progress. If you focus relentlessly on the need and what is really required to fully address it, these tough choices can be easier to make. In the case of sleep apnea, for example, we could have pressed forward with the sling, delivering results that were comparable to MADs; however, our goal—and the real need—was for a noninvasive solution that was substantially better, so we set it aside.

LEARNING FROM FAILURE

Innovation and entrepreneurship is an inherently risky undertaking, so you should expect to fail many times in your career. While this is never pleasant, it becomes a little easier if you think about failure as an opportunity to learn actively from your experience.

In the early stages of a project addressing problems with the prostate, we were looking for ways to more effectively treat benign prostatic hyperplasia (BPH). During concept generation, we came up with an idea that we called "capsular release." The concept was that by incising part of the capsule of the prostate, we might be able to relieve the pressure induced by the tissue compressing the prostatic urethra and thus relieve urinary symptoms. We got really excited about this idea, but when we went to the cadaver lab to test it, the concept did not work. Rather than simply giving up and moving on, we first refined and repeated our experiments to make sure the failure was not caused by flaws in our test designs. Then, we invested weeks examining the data to try advancing our understanding of what was actually happening. Through that investigation, we learned that the idea was tragically flawed; however, we discovered new characteristics of the prostate that we believed we could capitalize on to treat BPH in an entirely different way. By studying our failure rather than brushing it aside, we generated the insights that ultimately lead to the development of a novel technology and the company known today as NeoTract, Inc.

While spending so much time investigating our initial failure felt like a detour at the time, it ended up accelerating our path to a successful solution. That may not be the case every time, but there is always something to be learned from reflecting actively on your setbacks.

DO WHAT YOU CARE ABOUT

Nearly 30 years have passed since I filed my first patent application, yet the thrill of inventing something with the potential to positively affect the lives of thousands (or even hundreds of thousands) of patients is still a vivid memory. All of the inventions I have been a part of since then have been exciting. But I must admit that I find it especially rewarding when working on a problem that I am personally invested in solving. The best example of this is the story of the creation of Acclarent, which is now the Ear, Nose, and Throat division of Johnson & Johnson. I had chronic sinusitis throughout most of my life, but the condition had never been critical enough to warrant operative intervention. After attempting unsuccessfully to manage the sinusitis with everything from homeopathic remedies to antibiotics for many years, I decided to draw on my deep understanding of the need and the ExploraMed team's collective expertise in medical devices to explore whether there might be a more minimally-invasive way to solve the problem.

The key insight came to me several months into the study when looking at a computed tomography of a section of the sinus cavity. I could see all of the bones and how they were interdigitated in the sinuses—themselves a fascinating structure, as unique to an individual as a fingerprint—and I was reminded of the way coronary arteries look on an angiogram. In that moment, I wondered whether it would be possible to treat sinusitis with similar tools to those used for coronary interventions and angioplasty. Instead of using invasive tools that required removal of tissue and bone, could we navigate a balloon into the sinuses with a minimally-invasively approach and dilate the narrowed portion of the sinuses to allow more air and mucous to flow? That line of inquiry set off a chain of events that led to the creation of a new operative technique called "balloon sinuplasty" and to the formation of Acclarent (now the market leader in the field). A few years later, I had the procedure myself and felt particularly gratified that I, and thousands of patients like, me could finally get relief from sinusitis without the pain and high recurrence rate of traditional sinus surgery.

Clearly, not everyone will have the opportunity to cure a condition that they suffer from personally. But I urge you to choose a problem about which you are passionate. Getting any medical technology into patient care successfully is not an easy task, and it requires endless energy, astounding perseverance, and years and years of your life. But if you are working on a challenge that you feel strongly about, all of your blood, sweat, and tears will be far more rewarding.

STRENGTHS, WEAKNESSES, AND THE ART OF GETTING THE TEAM RIGHT

On any innovation project, it is essential to recognize and remain cognizant of your own expertise, talents, and weaknesses. With each new medical idea that I have pursued as an entrepreneur, I have stayed close to my strengths and have found team members who balance my weaknesses. My model involves the role of a key partner(s) who pairs my creativity with his or her knowledge, experience, and ability to execute. My open willingness to see no boundaries and consider the impossible becomes far more valuable when teamed with a partner's ability to be grounded and practical. The combination has been magical, and without these partners I would never have achieved any success.

Such was the case with NeoTract Inc., and also with Moximed Inc., where we produced the Kinespring and Atlas technology for the treatment of knee osteoarthritis. The success of both of these companies is rooted in the teams we assembled by bringing engineering, clinical, regulatory, management, and financial expertise to bear as a team rather than in any one person. None of these medical advancements could have been achieved without this multidisciplinary team approach. That same self-awareness and understanding of limitations and shortcomings is vital at subsequent stages of the innovation process as well, from company formation to product development to marketing. It is crucial to know what skills and expertise you have across your team, as well as those skills that you lack and will eventually require to bring your innovative idea to the patient.

PATIENCE AND PERSEVERANCE

If all of your needs, research, generation of the concept, and screening of potential solutions lead to the creation of a product prototype, that experience can feel like an extraordinary accomplishment on its own. For some, it may in fact be a culminating moment—not everyone opts to dive into the difficult work of product manufacture, marketing, and the ultimate planning and implementing of a new technology into patient care. This effort, which involves securing investment capital, protecting your intellectual property, navigating regulatory and reimbursement pathways, gathering compelling clinical evidence, defining a viable business model, and toppling many other potential barriers that can arise in bringing a product to market, is not for the faint of heart. Those who move forward must not only have a passion for addressing a particular clinical need but also must understand the extreme perseverance and patience needed to make a multiyear commitment to advancing their product.

With a rigorous innovation process to use as a roadmap, innovators can generally understand and act on the key activities, decisions, and risks involved in developing and commercializing a new health technology. Just be prepared that there is always a piece of your proposed innovation that at some point in time will be "unknowable." No analysis or data-driven approach can enable you to see around every corner.

In my experience, the important thing is to stay true to the need when you hit these blind spots. Also, recognize that your technology and the development and commercialization strategies that support it are almost certain to evolve (or shift dramatically!) over time as new information becomes available. Use your need as the guiding light but stay open to change in order to continue making progress. Sometimes it will be necessary to give up what you thought was fundamental in your product for what actually works, to really figure out what represents a win to you and your team, and even possibly to redefine what a win means. But if all of this is done in service of the need, the trade-offs you make to overcome hurdles on the path to market will prove worthwhile.

With Acclarent, we spent 5 years navigating unanticipated twists and turns establishing our technology in the market. For NeoTract, it took 12 years. But in both cases, we did not just develop, test, and launch new technologies; rather, we changed treatment paradigms, built markets, addressed risks ranging from regulation to reimbursement, and created sustainable businesses. It is not easy work, but it is how you make sure that your solution will deliver lasting benefits to patients.

THE ADVISOR EFFECT

As you seek to bring a technology forward, you will need undoubtedly to ask for help. My advice is to ask early and ask often. Even with a strong internal team in place, the advice and guidance of a robust network of external mentors and advisors is invaluable. Actively nurture a group of advisors with diverse backgrounds and points of view and seek their input regularly. In most MedTech start-ups, mentors play a main role in helping anticipate and mitigate the many risks mentioned above. You may not always agree with their input or take their advice, but you will benefit almost certainly from their guidance and learn something by engaging with them during your decision-making process.

Keep in mind that your investors may also be an important part of your network of mentors and advisors, as well as being your financial backing. As you think about raising funding, recognize that different investors can have wildly different views about their level of engagement in your company, as well as the length of the runway they give you. Many look to get a return on their investment in as little as 2–3 years, while, as noted, it can take a decade or longer to bring a new device or drug to market. Some investors may write a check but otherwise be passive in their involvement. Others will roll up their sleeves and be a resource at every step of the way.

Particularly if you are a first-time innovator, seek investors who are committed to helping you start and build your business over a reasonable and realistic period of time. That is how it was for us when we raised our first investment in ExploraMed from the team at New Enterprise Associates (NEA), one of the oldest and largest venture capital firms in the United States. Twenty years later, now as a General Partner leading the investments of the company in medical and health technologies, I hope to do the same for the people and companies that we back today and in the future.

DEFINING SUCCESS

Being an innovator in health technology is fundamentally about helping patients. As you navigate the innovation process, it is easy to begin thinking about success in terms of milestones reached or exits and transactions realized. Completing your clinical trial or signing a deal with an important strategic partner are both worthy triumphs, but these kinds of achievements are fleeting. For instance, no sooner do you obtain regulatory approval from the FDA than you must initiate the daunting task of pursuing reimbursement coverage from such groups as the Centers for Medicare and Medicaid Services (CMS) or private insurance companies; or, as you take a company public with great fanfare, you realize that the success of the stock will depend entirely on the growth of the company going forward, and that none of the prior accomplishments matter any longer. It can be a never-ending cycle. Even the attainment of a more terminal goal such as selling your company to a major MedTech corporation represents just a moment in time.

The point is that it can be helpful to think about success as more of a living, dynamic concept rather than the achievement of any given endpoint. Success is when you take a moment to pause, look around, and see a fantastic team of people, all working together and enjoying the challenges you are tackling cooperatively. It is the continual learning, growth, and development you realize as you overcome each new hurdle and celebrate each win. It is being able to pick yourself up after a major setback and recommit to your work with the team's fingers not pointed at each other but at how you all will collectively get the company back on track. But probably most importantly, it is the sense of deep accomplishment you get when physicians or patients tell you what a difference your technology has made for them and how it changed their lives in some meaningful way. To us, that is a much better gauge of your success as an innovator than any deal or exit event.

There are so many needs to solve and so many lives that can be improved with innovation—that is what compels us and our colleagues to seek new challenges and gain new ground with each project we initiate. Innovation by its very nature must continue, and so the cycle perpetuates itself from project to project and from one generation of innovators to the next. There really is no terminal endpoint when there is so much opportunity and so many needs to address. If you are a true innovator, you keep going.

I have relished the opportunity to make an impact through the various technologies I have been involved with, and by collaborating with my incredible cofounders, partners, and team members along the way. We hope that you too have the chance to make a difference through innovations in medical technology, and that the lives you touch inspire you to innovate again and again. As members of the medical community, innovation is in all of our DNA.

REFERENCES

[1] Makower J. Inspiration, perspiration, and execution: an innovator's perspective. Surgery 2017;161(5):1187–90. https://doi.org/10.1016/j.surg.2016.06.060.

[2] Yock P, Zenios S, Makower J, et al. Biodesign: the process of innovating medical technologies. 2nd ed. Cambridge, UK: Cambridge University Press; 2015.

Index

Note: Page numbers followed by *f* indicate figures and *t* indicate tables.

D

M

N

T

U

V

W

Y

Z

Printed in the United States
By Bookmasters